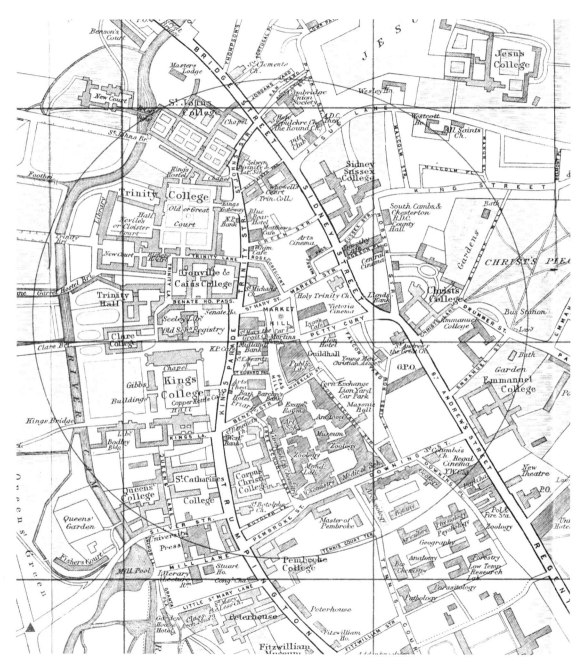

Cambridge city center, early 1950s. Detail of a map published by W. Heffer & Sons.

JAMES D. WATSON

THE ANNOTATED AND ILLUSTRATED
DOUBLE HELIX

Edited by

Alexander Gann & Jan Witkowski

SIMON & SCHUSTER

New York   London   Toronto   Sydney   New Delhi

Simon & Schuster
1230 Avenue of the Americas
New York, NY 10020

First Simon & Schuster hardcover edition November 2012

SIMON & SCHUSTER and colophon are registered
trademarks of Simon & Schuster, Inc.

For information about special discounts for bulk purchases,
please contact Simon & Schuster Special Sales at
1-866-506-1949 or business@simonandschuster.com.

The Simon & Schuster Speakers Bureau can bring authors
to your live event. For more information or to book an event,
contact the Simon & Schuster Speakers Bureau at
1-866-248-3049 or visit our website at www.simonspeakers.com.

Designed by Denise Weiss

Manufactured in the United States of America

10   9   8   7   6   5   4   3   2   1

Library of Congress Cataloging-in-Publication Data

Watson, James D., date.
    The annotated and illustrated double helix / James D. Watson ; edited by Alexander Gann &
Jan Witkowski. — 1st Simon & Schuster hardcover ed.
        p.   cm.
    Includes bibliographical references and index.
    I. Gann, Alexander.   II. Witkowski, J. A. (Jan Anthony), date.   III. Watson, James D., date.
Double helix.   IV. Title.
    [DNLM: 1. Watson, James D., date.   2. DNA.   3. Genetic Code.   4. Molecular Biology.
QU 58.5]
    572.8'6—dc23   2012037483
ISBN 978-1-4767-1549-0
ISBN 978-1-4767-1551-3 (ebook)

Portions of this book were first published in *Atlantic Monthly*.

For Naomi Mitchison

Contents

Preface to Annotated and Illustrated Edition

In Cold Spring Harbor's Blackford Bar, one evening in June 2010, Sydney Brenner suggested looking through the papers he had recently donated to the Cold Spring Harbor Laboratory Archives. Among his own papers were, he knew, some of Francis Crick's correspondence that had become muddled in with his during the 20 years they had shared an office in Cambridge. A few days later we discovered that the trove included letters to and from Crick written during the period when he and Jim Watson at Cambridge, and Maurice Wilkins and Rosalind Franklin in London, were searching for the structure of DNA.

Mislaid some 50 years earlier ("thrown out by an over efficient secretary," Crick believed), these letters had escaped the attention of the historians of molecular biology who first started looking into this new field in the mid-1960s. The letters provided some new insights into the proceedings, and in particular the personal relationships of the protagonists in the DNA story.

The most celebrated account of that story is *The Double Helix*, Watson's novelistic version of the events as they appeared to a 23-year-old American in Cambridge in the early 1950s. Written not in the tone of a formal autobiography nor in the measured language of the historian, his racy and thriller-like telling was reviled by some and praised by many upon publication in 1968.

In writing our article on the lost Crick correspondence, we naturally reread *The Double Helix*. We were struck by how Watson's account in the book accurately represented the vivid, contemporary descriptions of people and events found in the letters, and not just those of Crick and Wilkins, but Watson's own. The social whirl of parties, tennis, French lessons, holidays, and other events that featured prominently in the book—the "gossip," as Crick characterized it—were recorded in the weekly letters Watson wrote to his sister Elizabeth during his time in Cambridge. And the science covered in the book was also discussed in contemporary letters to Max Delbrück and other friends, and not just the DNA work, but Watson's research on bacterial genetics and tobacco mosaic virus, projects that figure prominently in the story. In all of this contemporary correspondence, the character of Watson himself—the brash, self-confident yet at times also self-deprecating, young man portrayed in his book—was transparent. We became intrigued to see all the contemporary accounts we could find—not just those revealed in the letters of Watson, Crick, and Wilkins, but of Franklin, Linus Pauling, and others as well.

We also noticed just how many other characters appear in *The Double Helix*—many unrelated to the central scientific story. Watson, eager to keep the narrative moving, often provides only the briefest of information, sometimes not even identifying the most intriguing of minor characters. We don't get to learn the interesting story of the "local doctor" who had rowing oars mounted on the wall of his surgery, or the identity of the "antiquarian architect" who kept his house free of gas and electricity—or anything much about Bertrand Fourcade except that he was the "most beautiful male" in Cambridge. And what was the novel of ill-judged sexual indiscretions of Cambridge dons that Watson reads at one point in the story? We wanted to know.

And so the idea of an annotated edition of *The Double Helix* took shape, a version in which an array of viewpoints and voices would be added as commentary, together with background information and illustrations to enrich the text. The current volume is the result. In addition to the numerous photographs (a number being published for the first time), we have reproduced many letters and other documents in full or in part as facsimiles. One of the pleasures of visiting archives is to see and handle original documents and while we cannot match that experience, we hope that readers will enjoy seeing letters and manuscripts as their original recipients saw them.

The sources of material used in our annotations are many, both published and unpublished. Of the former, we used many books—including the histories and biographies of the field. These are listed in the bibliography at the end of the book. Among unpublished sources, Watson's letters to his sister and also his parents are a major source of information about his Cambridge life and have not been used before, except by Watson, while his letters to Delbrück, Luria, and others provide scientific content. In addition to Watson's papers, we have drawn on those of Crick, Wilkins, Pauling, and Franklin among others. We have also included reminiscences written by Ray Gosling specially for this edition. Gosling worked with both Wilkins and Franklin in those years and actually took the most famous and influential diffraction pictures of DNA. The source for each annotation is included in a reference list at the back of the book.

In addition to the annotations and illustrations, we have added a number of other pieces. We have included Watson's account of winning the Nobel Prize, previously published in his later book, *Avoid Boring People.* This seems, on the occasion of the 50th anniversary of that award, a fitting conclusion to the tale. We have also added five appendices. These include one in which we reproduce facsimiles of the first letters Watson and Crick each wrote in 1953 describing the discovery, and another in which we publish for the first time a chapter from the draft manuscript of *The Double Helix* which was left out of the published book. While not describing anything new about the work on DNA, the missing chapter fills in the story of Watson's summer spent in the Alps in 1952.

We have corrected some errors of fact by adding annotations where necessary, but Watson's original text is unchanged.

It will be clear that this edition is not an exhaustive academic treatise. Rather, we chose items that appealed to us and hope that this somewhat quirky selection will prove useful and enjoyable to both new readers and those familiar with the original text.

Alexander Gann
Jan Witkowski
Cold Spring Harbor 2012

Sir Lawrence Bragg's Foreword
to the Original Edition

This account of the events which led to the solution of the structure of DNA, the fundamental genetical material, is unique in several ways. I was much pleased when Watson asked me to write the foreword.

There is in the first place its scientific interest. The discovery of the structure by Crick and Watson, with all its biological implications, has been one of the major scientific events of this century. The number of researches which it has inspired is amazing; it has caused an explosion in biochemistry which has transformed the science. I have been amongst those who have pressed the author to write his recollections while they are still fresh in his mind, knowing how important they would be as a contribution to the history of science. The result has exceeded expectation. The latter chapters, in which the birth of the new idea is described so vividly, are drama of the highest order; the tension mounts and mounts towards the final climax. I do not know of any other instance where one is able to share so intimately in the researcher's struggles and doubts and final triumph.

Then again, the story is a poignant example of a dilemma which may confront an investigator. He knows that a colleague has been working for years on a problem and has accumulated a mass of hard-won evidence, which has not yet been published because it is anticipated that success is just around the corner. He has seen this evidence and has good reason to believe that a method of attack which he can envisage, perhaps merely a new point of view, will lead straight to the solution. An offer of collaboration at such a stage might well be regarded as a trespass. Should he go ahead on his own? It is not easy to be sure whether the crucial new idea is really one's own or has been unconsciously assimilated in talks with others. The realization of this difficulty has led to the establishment of a somewhat vague code amongst scientists which recognizes a claim in a line of research staked out by a colleague—up to a certain point. When competition comes from more than one quarter, there is no need to hold back. This dilemma comes out clearly in the DNA story. It is a source of deep satisfaction to all intimately concerned that, in the award of the Nobel Prize in 1962, due recognition was given to the long, patient investigation by Wilkins at King's College (London) as well as to the brilliant and rapid final solution by Crick and Watson at Cambridge.

Finally, there is the human interest of the story—the impression made by Europe and England in particular upon a young man from the States. He writes with a Pepys-like frankness. Those who figure in the book must read it in a very forgiving spirit. One must remember that his book is not a history, but an autobiographical contribution to the history which will some day be written. As the author himself says, the book is a record of impressions rather than historical facts. The issues were often more complex, and the motives of those who had to deal with them were less tortuous, than he realized at the time. On the other hand, one must admit that his intuitive understanding of human frailty often strikes home.

The author has shown the manuscript to some of us who were involved in the story, and we have suggested corrections of historical fact here and there, but personally I have felt reluctant to alter too much because the freshness and directness with which impressions have been recorded is an essential part of the interest of this book.

W. L. B.

Sir Lawrence Bragg (1890–1971) was the director of the Cavendish Laboratory of Cambridge University at the time of the discovery of the double helix. He and his father, William Henry, the originators of X-ray crystallography, received the Nobel Prize in 1915.

Preface to the Original Edition

Here I relate my version of how the structure of DNA was discovered. In doing so I have tried to catch the atmosphere of the early postwar years in England, where most of the important events occurred. As I hope this book will show, science seldom proceeds in the straightforward logical manner imagined by outsiders. Instead, its steps forward (and sometimes backward) are often very human events in which personalities and cultural traditions play major roles. To this end I have attempted to re-create my first impressions of the relevant events and personalities rather than present an assessment which takes into account the many facts I have learned since the structure was found. Although the latter approach might be more objective, it would fail to convey the spirit of an adventure characterized both by youthful arrogance and by the belief that the truth, once found, would be simple as well as pretty. Thus many of the comments may seem one-sided and unfair, but this is often the case in the incomplete and hurried way in which human beings frequently decide to like or dislike a new idea or acquaintance. In any event, this account represents the way I saw things then, in 1951–1953: the ideas, the people, and myself.

I am aware that the other participants in this story would tell parts of it in other ways, sometimes because their memory of what happened differs from mine and, perhaps in even more cases, because no two people ever see the same events in exactly the same light. In this sense, no one will ever be able to write a definitive history of how the structure was established. Nonetheless, I feel the story should be told, partly because many of my scientific friends have expressed curiosity about how the double helix was found, and to them an incomplete version is better than none. But even more important, I believe, there remains general ignorance about how science is "done." That is not to say that all science is done in the manner described here. This is far from the case, for styles of scientific research vary almost as much as human personalities. On the other hand, I do not believe that the way DNA came out constitutes an odd exception to a scientific world complicated by the contradictory pulls of ambition and the sense of fair play.

The thought that I should write this book has been with me almost from the moment the double helix was found. Thus my memory of many of the significant events is much more complete than that of most other episodes in my life. I also have made extensive use of letters written at virtually weekly intervals to my parents. These were especially helpful in exactly dating a number of the incidents. Equally important have been the valuable comments by various friends who kindly read earlier versions and gave in some instances quite detailed accounts of incidents that I had referred to in less complete form. To be sure, there are cases where my recollections differ from theirs, and so this book must be regarded as my view of the matter.

Some of the earlier chapters were written in the homes of Albert Szent-Györgyi, John A. Wheeler, and John Cairns, and I wish to thank them for quiet rooms with tables overlooking the ocean. The later chapters were written with the help of a Guggenheim Fellowship, which allowed me to return briefly to the other Cambridge and the kind hospitality of the Provost and Fellows of King's College.

As far as possible I have included photographs taken at the time the story occurred, and in particular I want to thank Herbert Gutfreund, Peter Pauling, Hugh Huxley, and Gunther Stent for sending me some of their snapshots. For editorial assistance I'm much indebted to Libby Aldrich for the quick, perceptive remarks expected from our best Radcliffe students and to Joyce Lebowitz both for keeping me from completely misusing the English language and for innumerable comments about what a good book must do. Finally, I wish to express thanks for the immense help Thomas J. Wilson has given me from the time he saw the first draft. Without his wise, warm, and sensible advice, the appearance of this book, in what I hope is the right form, might never have occurred.

J. D. W.
Harvard University
Cambridge, Massachusetts
November 1967

THE ANNOTATED AND ILLUSTRATED

DOUBLE HELIX

Prologue from the Original Edition

In the summer of 1955, I arranged to join some friends who were going into the Alps. Alfred Tissieres, then a Fellow at King's, had said he would get me to the top of the Rothorn, and even though I panic at voids this did not seem to be the time to be a coward. So after getting in shape by letting a guide lead me up the Allinin, I took the two-hour postal-bus trip to Zinal, hoping that the driver was not carsick as he lurched the bus around the narrow road twisting above the falling rock slopes. Then I saw Alfred standing in front of the hotel, talking with a long-mustached Trinity don who had been in India during the war.

Since Alfred was still out of training, we decided to spend the afternoon walking up to a small restaurant which lay at the base of the huge glacier falling down off the Obergabelhorn and over which we were to walk the next day. We were only a few minutes out of sight of the hotel when we saw a party coming down upon us, and I quickly recognized one of the climbers. He was Willy Seeds, a scientist who several years before had worked at King's College, London, with Maurice Wilkins on the optical properties of DNA fibers. Willy soon spotted me, slowed down, and momentarily gave the impression that he might remove his rucksack and chat for a while. But all he said was, "How's Honest Jim?" and quickly increasing his pace was soon below me on the path.[1]

Later as I trudged upward, I thought again about our earlier meetings in London. Then DNA was still a mystery, up for grabs, and no one was sure who would get it and whether he would deserve it if it proved as exciting as we semisecretly believed. But now the race was over and, as one of the winners, I knew the tale was not simple and certainly not as the newspapers reported. Chiefly it was a matter of five people: Maurice Wilkins, Rosalind Franklin, Linus Pauling, Francis Crick, and me. And as Francis was the dominant force in shaping my part, I will start the story with him.

[1] Willy Seeds' remark provided Watson with the title he originally wanted to use for what became *The Double Helix*. See the handwritten title page from an early draft (shown overleaf) and Appendix 4.

1

HONEST JIM

(A discription of a very great discovery)

by

J.D. Wat—

Chapter 1

I have never seen Francis Crick in a modest mood. Perhaps in other company he is that way, but I have never had reason so to judge him. It has nothing to do with his present fame. Already he is much talked about, usually with reverence, and someday he may be considered in the category of Rutherford or Bohr. But this was not true when, in the fall of 1951, I came to the Cavendish Laboratory of Cambridge University to join a small group of physicists and chemists working on the three-dimensional structures of proteins.[1] At that time he was thirty-five, yet almost totally unknown. Although some of his closest colleagues realized the value of his quick, penetrating mind and frequently sought his advice, he was often not appreciated, and most people thought he talked too much.

[1] The Cavendish Laboratory was founded in 1874 through the gift of William Cavendish, Seventh Duke of Devonshire. The first Cavendish Professor of Experimental Physics was James Clerk Maxwell, and other notable Cavendish Professors include Nobel Laureates Lord Rayleigh, J. J. Thomson, Lord Rutherford, Lawrence Bragg, and Nevill Mott.

The Cavendish Laboratory, Free School Lane, Cambridge, 1940s.

Max Perutz, 1950s.

[2] Lawrence Bragg together with his father, William Henry Bragg, showed how X-ray diffraction patterns could be used to deduce the atomic structures of crystals. The Braggs won the Nobel Prize for Physics in 1915, the only father–son pair to share a Nobel Prize, and Lawrence Bragg, only 25 when the award was made, remains the youngest ever winner. He was serving in the trenches in the Great War when he learned he had won the Prize.

Leading the unit to which Francis belonged was Max Perutz, an Austrian-born chemist who came to England in 1936. He had been collecting X-ray diffraction data from hemoglobin crystals for over ten years and was just beginning to get somewhere. Helping him was Sir Lawrence Bragg, the director of the Cavendish. For almost forty years Bragg, a Nobel Prize winner and one of the founders of crystallography, had been watching X-ray diffraction methods solve structures of ever-increasing difficulty. The more complex the molecule, the happier Bragg became when a new method allowed its elucidation.[2] Thus in the immediate postwar years he was especially keen about the possibility of solving the structures of proteins, the most complicated of all molecules. Often, when administrative duties permitted, he visited Perutz' office to discuss recently accumulated X-ray data. Then he would return home to see if he could interpret them.

Somewhere between Bragg the theorist and Perutz the experimentalist was Francis, who occasionally did experiments but more often was immersed in the theories for solving protein structures. Often he came up with something novel, would become enormously excited, and immediately tell it to anyone who would listen. A day or so later he would often realize that his theory did not work and return to experiments, until boredom generated a new attack on theory.

There was much drama connected with these ideas. They did a great deal to liven up the at-

William Lawrence Bragg with his father, William Henry Bragg, 1930s.

mosphere of the lab, where experiments usually lasted several months to years. This came partly from the volume of Crick's voice: he talked louder and faster than anyone else and, when he laughed, his location within the Cavendish was obvious. Almost everyone enjoyed these manic moments, especially when we had the time to listen attentively and to tell him bluntly when we lost the train of his argument. But there was one notable exception. Conversations with Crick frequently upset Sir Lawrence Bragg, and the sound of his voice was often sufficient to make Bragg move to a safer room. Only infrequently would he come to tea in the Cavendish, since it meant enduring Crick's booming over the tea room.[3] Even then Bragg was not completely safe. On two occasions the corridor outside his office was flooded with water pouring out of a laboratory in which Crick was working. Francis, with his interest in theory, had neglected to fasten securely the rubber tubing around his suction pump.

At the time of my arrival, Francis' theories spread far beyond the confines of protein crystallography. Anything important would attract him, and he frequently visited other labs to see which new experiments had been done.[4]

[3] Morning and afternoon tea, rituals of laboratory life in British academic institutions, allowed members of a laboratory or department to assemble and talk over tea and, with luck, biscuits (cookies). However, teatime often reinforced class distinctions, with staff scientists in a comfortably furnished room, while technicians, secretaries, and graduate students had to be content with less desirable surroundings.

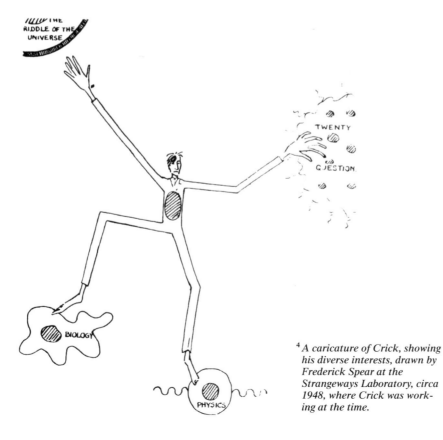

Francis next to a Cavendish X-ray tube, early 1950s.

[4] *A caricature of Crick, showing his diverse interests, drawn by Frederick Spear at the Strangeways Laboratory, circa 1948, where Crick was working at the time.*

Though he was generally polite and considerate of colleagues who did not realize the real meaning of their latest experiments, he would never hide this fact from them. Almost immediately he would suggest a rash of new experiments that should confirm his interpretation. Moreover, he could not refrain from subsequently telling all who would listen how his clever new idea might set science ahead.

As a result, there existed an unspoken yet real fear of Crick, especially among his contemporaries who had yet to establish their reputations. The quick manner in which he seized their facts and tried to reduce them to coherent patterns frequently made his friends' stomachs sink with the apprehension that, all too often in the near future, he would succeed, and expose to the world the fuzziness of minds hidden from direct view by the considerate, well-spoken manners of the Cambridge colleges.

Though he had dining rights for one meal a week at Caius College, he was not yet a fellow of any college. Partly this was his own choice. Clearly he did not want to be burdened by the

The inner quad of Gonville and Caius College, Cambridge.

unnecessary sight of undergraduate tutees.[5] Also a factor was his laugh, against which many dons would most certainly rebel if subjected to its shattering bang more than once a week. I am sure this occasionally bothered Francis, even though he obviously knew that most High Table life is dominated by pedantic, middle-aged men incapable of either amusing or educating him in anything worthwhile. There always existed King's College, opulently nonconformist and clearly capable of absorbing him without any loss of his or its character.[6] But despite much effort on the part of his friends, who knew he was a delightful dinner companion, they were never able to hide the fact that a stray remark over sherry might bring Francis smack into your life.

[5] The fellows of a college are responsible for its administration. Had he been a fellow, Crick would have had more privileges than just dining rights for one meal a week, but those privileges did not come for free: fellows typically had to give tutorial classes to undergraduates.

[6] The current opulence of King's College, Cambridge is largely due to John Maynard Keynes, in part because of his financial acumen as bursar of the College and through bequests he made in his will. Paradoxically, the College was regarded as left-wing because its fellows and students were noted for their socialist and communist sympathies.

Chapter 2

[1] *What Is Life?* is based on a series of lectures given by Erwin Schrödinger at Trinity College, Dublin, in 1944. Although Crick, Maurice Wilkins, Erwin Chargaff, and Watson himself were influenced by *What Is Life?*, others were less impressed. Perutz later wrote: "Sadly…a close study of his book and of the original related literature has shown me that what was true in his book was not original, and most of what was original was known not to be true even when the book was written."

Crick wrote to Schrödinger on August 12, 1953, after the discovery of the double helix, sending reprints and telling Schrödinger that both he and Watson had been influenced by *What Is Life?*.

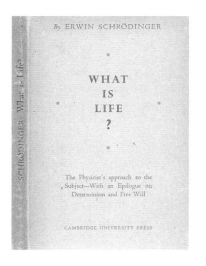

Before my arrival in Cambridge, Francis only occasionally thought about deoxyribonucleic acid (DNA) and its role in heredity. This was not because he thought it uninteresting. Quite the contrary. A major factor in his leaving physics and developing an interest in biology had been the reading in 1946 of *What Is Life?* by the noted theoretical physicist Erwin Schrödinger.[1] This book

Erwin Schrödinger, 1926.

very elegantly propounded the belief that genes were the key components of living cells and that, to understand what life is, we must know how genes act. When Schrödinger wrote his book (1944), there was general acceptance that genes were special types of protein molecules. But almost at this same time the bacteriologist O. T. Avery was carrying out experiments at the Rockefeller Institute in New York which showed that hereditary traits could be transmitted from one bacterial cell to another by purified DNA molecules.[2]

Given the fact that DNA was known to occur in the chromosomes of all cells, Avery's experiments strongly suggested that future experiments would show that all genes

Oswald T. Avery, 1920s.

were composed of DNA. If true, this meant to Francis that proteins would not be the Rosetta Stone for unraveling the true secret of life. Instead, DNA would have to provide the key to enable us to find out how the genes determined, among other characteristics, the color of our hair, our eyes, most likely our comparative intelligence, and maybe even our potential to amuse others.

If we are right—& of course that's not yet proven, then it means that nucleic acids are not merely structurally important but functionally active substances in determining the biochemical activities and specific characteristics of cells—& that by means of a known chemical substance it is possible to induce predictable *and* hereditary *changes in cells. This is something that has long been the dream of geneticists…But with mechanisms I am not now concerned—one step at a time—& the first step is, what is the chemical nature of the transforming principle? Some one else can work out the rest— of course the problem bristles with implications…It touches biochemistry…It touches genetics, enzyme chemistry, cell metabolism & carbohydrate synthesis—etc—But lately it takes a lot of well documented evidence to convince anyone that the sodium salt of deoxyribose nucleic acid, protein free, could possibly be endowed with such biologically active & specific properties, & that evidence we are now trying to get. It's lots of fun to blow bubbles—but it's wiser to prick them yourself before someone else tries to.*

[2]*…Avery was cautious in the published paper about claiming that DNA was the genetic material. He was more confident in this letter to his brother, Roy (May 13, 1943).*

Of course there were scientists who thought the evidence favoring DNA was inconclusive and preferred to believe that genes were protein molecules. Francis, however, did not worry about these skeptics. Many were cantankerous fools who unfailingly backed the wrong horses. One could not be a successful scientist without realizing that, in contrast to the popular conception supported by newspapers and mothers of scientists, a goodly number of scientists are not only narrow-minded and dull, but also just stupid.

Francis, nonetheless, was not then prepared to jump into the DNA world. Its basic importance did not seem sufficient cause by itself to lead him out of the protein field which he had worked in only two years and was just beginning to master intellectually. In addition, his colleagues at the Cavendish were only marginally interested in the nucleic acids, and even in the best of financial circumstances it would take two or three years to set up a new research group primarily devoted to using X rays to look at the DNA structure.

King's College, London, 1950.

Moreover, such a decision would create an awkward personal situation. At this time molecular work on DNA in England was, for all practical purposes, the per-

Maurice Wilkins, 1958.

sonal property of Maurice Wilkins, a bachelor who worked in London at King's College.[3] Like Francis, Maurice had been a physicist and also used X-ray diffraction as his principal tool of research. It would have looked very bad if Francis had jumped in on a problem that Maurice had worked over for several years. The matter was even worse because the two, almost equal in age, knew each other and, before Francis remarried, had frequently met for lunch or dinner to talk about science.

It would have been much easier if they had been living in different countries. The combination of England's coziness—all the important people, if not related by marriage, seemed to know one another—plus the English

[3] Not to be confused with King's College, Cambridge, King's College, London, was established in 1829 to provide a religious environment for education in response to the founding in 1826 of the secular "London University" (later University College, London) which admitted non-Anglican Christians, Jews, and utilitarians. In keeping with its religious origins, King's College has a large baroque chapel within its main building.

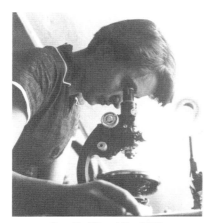

Rosalind Franklin, 1955.

[4] While Watson writes that Franklin worked in Wilkins' laboratory and was to be his assistant, she had been hired by Randall to take charge of the DNA project, as the letter opposite makes clear.

[5] The nickname "Rosy" (often spelled "Rosie" in correspondence of the time) was bestowed on Franklin by William Seeds who was the joker in the Biophysics Unit. This is the same "Willy" Seeds who coined the phrase "Honest Jim" as described in the Preface. Wilkins' nickname was "Uncle."

sense of fair play would not allow Francis to move in on Maurice's problem. In France, where fair play obviously did not exist, these problems would not have arisen. The States also would not have permitted such a situation to develop. One would not expect someone at Berkeley to ignore a first-rate problem merely because someone at Cal Tech had started first. In England, however, it simply would not look right.

Even worse, Maurice continually frustrated Francis by never seeming enthusiastic enough about DNA. He appeared to enjoy slowly understating important arguments. It was not a question of intelligence or common sense. Maurice clearly had both; witness his seizing DNA before almost everyone else. It was that Francis felt he could never get the message over to Maurice that you did not move cautiously when you were holding dynamite like DNA. Moreover, it was increasingly difficult to take Maurice's mind off his assistant, Rosalind Franklin.[4]

Not that he was at all in love with Rosy, as we called her from a distance.[5] Just the opposite—almost from the moment she arrived in Maurice's lab, they began to upset each other. Maurice, a beginner in X-ray diffraction work, wanted some professional help and hoped that Rosy, a trained crystallographer, could speed up his research. Rosy, however, did not see the situation this way. She claimed that she had been given DNA for her own problem and would not think of herself as Maurice's assistant.[6]

I suspect that in the beginning Maurice hoped that Rosy would calm down. Yet mere inspection suggested that she would not easily bend. By choice she did not emphasize her feminine qualities. Though her features were strong, she was not unattractive and might have been quite stunning had she taken even a mild interest in clothes. This she did not. There was never lipstick to contrast with her straight black hair, while at the age of thirty-one her dresses showed all the imagination of English blue-stocking adolescents. So it was quite easy to imagine her the product of an unsatisfied mother who unduly stressed the desirability of professional careers that could save bright girls from marriages to dull men. But this was not the case. Her dedicated, austere life could not be thus explained—she was the daughter of a solidly comfortable, erudite banking family.

UNIVERSITY OF LONDON KING'S COLLEGE.

TEMPLE BAR 5651
(4 LINES)

From The Wheatstone Professor of Physics,
J. T. RANDALL, F.R.S.

STRAND. W.C.2.

Dr. R. Franklin,
12 quai Henri IV,
Paris IV.

4th December, 1950

Dear Dr. Franklin,

I am sorry I have taken so long to reply to your letter of November 24th. The real difficulty has been that the X-ray work here is in a somewhat fluid state and the slant on the research has changed rather since you were last yere.

After very careful consideration and discussion with the senior people concerned, it now seems that it would be a good deal more important for you to investigate the structure of certain biological fibres in which we are interested, both by low and high angle diffraction, rather than to continue with the original project of work on solutions as the major one.

Dr. Stokes, as I have long inferred, really wishes to concern himself almost entirely with theoretical problems in the future and these will not necessarily be confined to X-ray optics. It will probably involve microscopy in general. This means that as far as the experimental X-ray effort is concerned there will be at the moment only yourself and Gosling, together with the temporary assistance of a graduate from Syracuse, Mrs. Heller. Gosling, working in conjunction with Wilkins, has already found that fibres of desoxyribose nucleic acid derived from material provided by Professor Signer of Bern gives remarkably good fibre diagrams. The fibres are strongly negatively birefringent and become positive on stretching, and are reversible in a moist atmosphere. As you no doubt know, nucleic acid is an extremely important constituent of cells and it seems to us that it would be very valuable if this could be followed up in detail. If you are agreeable to this change of plan it would seem that there is no necessity immediately to design a camera for work on solutions. The camera will, however, be extremely valuable in searching for large spacings from such fibres.

I hope you will understand that I am not in this way suggesting that we should give up all thought of work on solutions, but we do feel that the work on fibres would be more immediately profitable and, perhaps, fundamental.

I think I must leave to you the question as to whether you come over here for a day or two to discuss these matters further. It now seems so near to the time when you will actually be working here that it is

perhaps hardly necessary for you to make the special journey. On the other hand there may be things which you could organize on the apparatus side in Paris and you could hardly do this without further discussion with us. The change of programme, such as I have suggested, will probably mean that we should obtain the formal consent of the Fellowship Committee; there is no hurry about this and there is no doubt about the answer.

Dr. Price has just heard from Mr. Heins of the Rockefeller Foundation that orders have now been placed for your apparatus.

Yours sincerely,

J T Randall

[6] Franklin's claim that DNA was her project is borne out by this letter to her from John Randall, her future boss. Dated December 4, 1950 and written before she arrived at King's College, London, it is, as Brenda Maddox vividly described it, "…artful and ambiguous…packed with half-truths and buried meanings that would explode in Rosalind's face before too long and, not incidentally, alter the course of scientific history."

Randall tells Franklin that she will be working on DNA rather than on proteins in solution, a change that Wilkins had urged on Randall. Randall also told her that only she and Raymond Gosling, a Ph.D. student at the time with Wilkins, would be working on DNA. Randall had not discussed this with Wilkins although the letter implies that he had.

Wilkins, who did not know of this letter until many years later, was away on holiday when Franklin arrived at King's, and so missed a meeting where Gosling was formally assigned to Franklin. Wilkins wrote about the letter in his memoir published in 2003: "My opinion is very clear: Randall was very wrong to have written to Rosalind telling her that Stokes and I wished to stop our X-ray work on DNA, without consulting us. After Raymond and I got a clear crystal-line X-ray pattern I was very eager to continue that work, and on holiday had decided that I must continue full-time on DNA X-ray work and stop all other research. If Randall really believed the idea that I did not wish to continue my DNA X-ray research, he had deceived himself—perhaps because *he* so strongly wanted to be involved in DNA research."

[7] While the combination rooms (common rooms) at King's were separate for men and women, this mysogynist attitude did not extend to Randall's department. Raymond Gosling commented in a 2010 radio interview that "The labs that Randall ran were full of women." Horace Freeland Judson found that of the 31 scientists listed as being in the Biophysics Unit in December 1952, eight were women, and Honor Fell served as the Unit's Senior Biological Advisor.

It wasn't just King's that Franklin found unappealing at this time. After Paris, returning to England seemed gloomy in many ways—from the weather to the drab postwar look and feel of the place. In a letter to her friend Anne Sayre on March 1, 1952, she asked: "What is it that makes one's home country seem so awful after returning from living in a foreign country one has come to love?...To put it at its lowest, I suspect that I enjoyed being a more interesting person in France than I am in England..."

Clearly Rosy had to go or be put in her place. The former was obviously preferable because, given her belligerent moods, it would be very difficult for Maurice to maintain a dominant position that would allow him to think unhindered about DNA. Not that at times he didn't see some reason for her complaints—King's had two combination rooms, one for men, the other for women, certainly a thing of the past.[7] But he was not responsible, and it was no pleasure to bear the cross for the added barb that the women's combination room remained dingily pokey whereas money had been spent to make life agreeable for him and his friends when they had their morning coffee.

Unfortunately, Maurice could not see any decent way to give Rosy the boot. To start with, she had been given to think that she had a position for several years. Also, there was no denying she had a good brain. If she could only keep her emotions under control, there would be a good chance that she could really help him. But merely wishing for relations to improve was taking something of a gamble, for Cal Tech's fabulous chemist Linus Pauling was not subject to the confines of British fair play. Sooner or later Linus, who had just turned fifty, was bound to try for the most important of all scientific prizes. There was no doubt that he was interested. Our first principles told us that Pauling could not be the greatest of all chemists without realizing that DNA was the most golden of all molecules. Moreover, there was definite proof. Maurice had received a letter from Linus asking for a copy of the crystalline DNA X-ray photographs. After some hesitation he wrote back saying that he wanted to look more closely at the data before releasing the pictures.[8]

Linus Pauling examining a crystal, 1947.

UNIVERSITY OF LONDON KING'S COLLEGE.

From The Wheatstone Professor of Physics.

TEMple Bar 5651
(o lines).

J. T. RANDALL, F.R.S.

STRAND. W.C.2.

Professor Linus Pauling,
California Institute of Technology,
Gates and Crellin Laboratories of Chemistry,
Pasadena 4.

28th August, 1951

Dear Professor Pauling,

It was nice to hear from you on returning from holiday yesterday.

I am sorry that Oster is rather misinformed about our intentions with regard to nucleic acid. Wilkins and others are busily engaged in working out the interpretation of the desoxyribosenucleic acid X-ray photographs and it would not be fair to them, or to the efforts of the laboratory as a whole, to hand these over to you. Wilkins has, of course, obtained a good deal of information already from his optical studies and it is natural that he should wish to carry through the X-ray investigations.

I was not able to attend the Gordon Conference which opened at New Hampton yesterday but Wilkins is attending and will be talking about his optical work.

With very best wishes,

Yours sincerely,

Randall's letter to Pauling, August 28, 1951.

[8] Gerald Oster, at the Polytechnic Institute of Brooklyn, had told Pauling: "I hope you'll write to Prof. J. T. Randall, King's College, Strand, London. His coworker, Dr. M. Wilkins, told me he had some good fibre pictures of nucleic acid" (August 9, 1951). Wilkins passed the resulting letter to Randall, and it was Randall himself who rebuffed Pauling in the letter shown here of August 28, 1951. Pauling responded to Randall: "It is quite clear that Dr. Oster was misinformed when he spoke to me—he said quite definitely that Wilkins was not planning to carry out an interpretation of the X-ray photographs. I was, of course, surprised at this, but it seemed to me worth while to follow up his suggestion that I write to you."

All this was most unsettling to Maurice. He had not escaped into biology only to find it personally as objectionable as physics, with its atomic consequences.[9] The combination of both Linus and Francis breathing down his neck often made it very difficult to sleep. But at least Pauling was six thousand miles away, and even Francis was separated by a two-hour rail journey. The real problem, then, was Rosy. The thought could not be avoided that the best home for a feminist was in another person's lab.

[9] "…as objectionable as physics, with its atomic consequences" is a reference to the Manhattan Project which, between 1942 and 1946, employed many hundreds of physicists and other scientists. Wilkins had worked at the University of California, Berkeley, on methods to prepare uranium-235 (used in the atomic bombs) from natural uranium. Following the war, Wilkins, like Leo Szilard and others, felt moral revulsion at the applications of science to warfare, and moved from physics to biology.

The uranium separation project was run by the Australian physicist Harrie Massey who advised Wilkins to read Schrödinger's *What Is Life?* Crick also knew Massey who, before his assignment to Berkeley, headed the Mine Design Department. According to Watson, it was also Massey who introduced Crick to *What Is Life?*

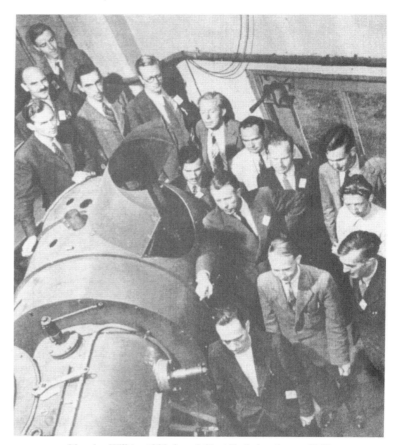

Maurice Wilkins (fifth from left) at Berkeley, August 1945.

Chapter 3

It was Wilkins who had first excited me about X-ray work on DNA. This happened at Naples when a small scientific meeting was held on the structures of the large molecules found in living cells.[1] Then it was the spring of 1951, before I knew of Francis Crick's existence. Already I was much involved with DNA, since I was in Europe on a postdoctoral fellowship to learn its biochemistry. My interest in DNA had grown out of a desire, first picked up while a senior in college, to learn what the gene was. Later, in graduate school at Indiana University, it was my hope that the gene might be solved without my learning any chemistry.[2] This wish partially arose from laziness since, as an undergraduate at the University of Chicago, I was principally interested in birds and managed to avoid taking any chemistry or physics courses which looked of even

J. D. Watson, Indiana University, late 1940s.

Watson (3rd from left) and friends birdwatching, 1946.

15

medium difficulty. Briefly the Indiana biochemists encouraged me to learn organic chemistry, but after I used a bunsen burner to warm up some benzene, I was relieved from further true chemistry. It was safer to turn out an uneducated Ph.D. than to risk another explosion.

So I was not faced with the prospect of absorbing chemistry until I went to Copenhagen to do my postdoctoral research with the biochemist Herman Kalckar. Journeying abroad initially appeared the perfect solution to the complete lack of chemical facts in my head, a condition at times encouraged by my Ph.D. supervisor, the Italian-trained microbiologist Salvador Luria.[3] He positively abhorred most chemists, especially the competitive variety out of the jungles of New York City. Kalckar, however, was obviously cultivated, and Luria hoped that in his civilized, continental company I would learn the necessary tools to do chemical research, without needing to react against the profit-oriented organic chemists.

Then Luria's experiments largely dealt with the multiplication of bacterial viruses (bacteriophages, or phages for short). For some years the suspicion had existed among the more inspired geneticists that viruses were a form of naked genes. If so, the best way to find out what a gene was and how it duplicated was

[3] Watson had decided Hermann Muller's research on fruit fly genetics was old-fashioned and turned instead for his Ph.D. project to Salvador Luria's cutting-edge research.

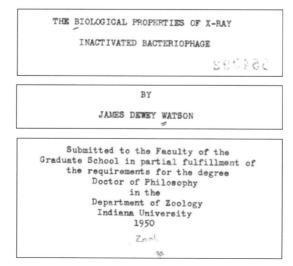

THE BIOLOGICAL PROPERTIES OF X-RAY

INACTIVATED BACTERIOPHAGE

BY

JAMES DEWEY WATSON

Submitted to the Faculty of the
Graduate School in partial fulfillment of
the requirements for the degree
Doctor of Philosophy
in the
Department of Zoology
Indiana University
1950

The title page of Watson's Ph.D. thesis. Even at the time, Watson's Ph.D. research topic was, he thought, boring.

Max Delbrück (standing) and Salvador Luria looking at phage plaques at Cold Spring Harbor, 1941.

to study the properties of viruses. Thus, as the simplest viruses were the phages, there had sprung up between 1940 and 1950 a growing number of scientists (the phage group) who studied phages with the hope that they would eventually learn how the genes controlled cellular heredity. Leading this group were Luria and his German-born friend, the theoretical physicist Max Delbrück, then a professor at Cal Tech.[4] While Delbrück kept hoping that purely genetic tricks could solve the problem, Luria more often wondered whether the real answer would come only after the chemical structure of a virus (gene) had been cracked open. Deep down he knew that it is impossible to describe the behavior of something when you don't know what it is. Thus, knowing he could never bring himself to learn chemistry, Luria felt the wisest course was to send me, his first serious student, to a chemist.

    He had no difficulty deciding between a protein chemist and a nucleic-acid chemist. Though only about one half the mass of a bacterial virus was DNA (the other half being protein), Avery's experiment made it smell like the essential ge-

[4] Luria and Delbrück met in Philadelphia in December 1940. They began working together on phage in the summer of 1941 at Cold Spring Harbor. In the following year, they began the Phage Course there, teaching the techniques of phage research and establishing a critical style of research which came to characterize members of the phage group.

netic material. So working out DNA's chemical structure might be the essential step in learning how genes duplicated. Nonetheless, in contrast to the proteins, the solid chemical facts known about DNA were meager. Only a few chemists

Herman Kalckar at the Phage Course, 1945, annotation by Manny Delbrück.

worked with it and, except for the fact that nucleic acids were very large molecules built up from smaller building blocks, the nucleotides, there was almost nothing chemical that the geneticist could grasp at. Moreover, the chemists who did work on DNA were almost always organic chemists with no interest in genetics. Kalckar was a bright exception. In the summer of 1945 he had come to the lab at Cold Spring Harbor, New York, to take Delbrück's course on bacterial viruses.[5] Thus both Luria and Delbrück hoped the Copenhagen lab would be the place where the combined techniques of chemistry and genetics might eventually yield real biological dividends.

Their plan, however, was a complete flop. Herman did not stimulate me in the slightest. I found myself just as indifferent to nucleic-acid chemistry

[5] There were two institutes at Cold Spring Harbor when Watson first visited: the Biological Laboratory (established in 1890) and the Department of Genetics of the Carnegie Institute of Washington (established in 1912 as the Station for Experimental Evolution). Milislav Demerec was director of both institutes. Nobel Laureates Al Hershey and Barbara McClintock were on the staff of the Department of Genetics.

Cold Spring Harbor Laboratory.

Snapshot taken at the microbial genetics meeting, held at the Institute for Theoretical Physics, Copenhagen, March 1951. Here Niels Bohr, one of the giants of 20th century physics, poses with Watson and other participants in the meeting. First row: Ole Maaløe, R. Latarjet, E. Wollman. Second row: N. Bohr, N. Visconti, G. Ehrensvaard, Wolf Weidel, H. Hyden, V. Bonifas, Gunther Stent, Herman Kalckar, Barbara Wright, Jim Watson, M. Westergaard.

in his lab as I had been in the States. This was partly because I could not see how the type of problem on which he was then working (the metabolism of nucleotides) would lead to anything of immediate interest to genetics. There was also the fact that, though Herman was obviously civilized, it was impossible to understand him.

I was able, however, to follow the English of Herman's close friend Ole Maaløe. Ole had just returned from the States (Cal Tech), where he had become very excited about the same phages on which I had worked for my degree. Upon his return he gave up his previous research problem and was devoting full time to phage. Then he was the only Dane working with phage and so was quite pleased that I and Gunther Stent, a phage worker from Delbrück's lab, had come to do research with Herman. Soon Gunther and I found

[6] Watson had applied to the National Research Council (NRC) for a fellowship supported by Merck & Co. The Merck Fellowship Board approved his application on March 18, 1950 and he was awarded a stipend of $3000 and travel expenses of $500. On this occasion, Watson's fellowship was renewed but he later ran afoul of the Merck Fellowship Board as described in Appendix 3.

[7] Kalckar had separated from his wife Vibeke "Vips" Meyer and taken up with Barbara Wright, a postdoc in his laboratory, whom he later married. (Wright is shown in the picture on page 19.) Watson had met her earlier at Caltech, in 1949. There he, Wright, Gunther Stent, and Wolf Weidel had gone camping, an adventure that resulted in the short-lived arrest of Watson and Wright by Catalina Island police.

Watson wrote a long letter to Delbrück (March 22, 1951) about the Kalckar affair and the effect it had on the mood of the laboratory: "I find it difficult to describe the very morbid feeling which pervaded Herman's lab during this period."

ourselves going regularly to visit Ole's lab, located several miles from Herman's, and within several weeks we were both actively doing experiments with Ole.

At first I occasionally felt ill at ease doing conventional phage work with Ole, since my fellowship was explicitly awarded to enable me to learn biochemistry with Herman; in a strictly literal sense I was violating its terms. Moreover, less than three months after my arrival in Copenhagen I was asked to propose plans for the following year. This was no simple matter, for I had no plans. The only safe course was to ask for funds to spend another year with Herman. It would have been risky to say that I could not make myself enjoy biochemistry. Furthermore, I could see no reason why they should not permit me to change my plans after the renewal was granted. I thus wrote to Washington saying that I wished to remain in the stimulating environment of Copenhagen. As expected, my fellowship was then renewed. It made sense to let Kalckar (whom several of the fellowship electors knew personally) train another biochemist.[6]

There was also the question of Herman's feelings. Perhaps he minded the fact that I was only seldom around. True, he appeared very vague about most things and might not yet have really noticed. Fortunately, however, these fears never had time to develop seriously. Through a completely unanticipated event my moral conscience became clear. One day early in December, I cycled over to Herman's lab expecting another charming yet totally incomprehensible conversation. This time, however, I found Herman could be understood. He had something important to let out: his marriage was over, and he hoped to obtain a divorce. This fact was soon no secret—everyone else in the lab was also told. Within a few days it became apparent that Herman's mind was not going to concentrate on science for some time, for perhaps as long as I would remain in Copenhagen. So the fact that he did not have to teach me nucleic-acid biochemistry was obviously a godsend. I could cycle each day over to Ole's lab, knowing it was clearly better to deceive the fellowship electors about where I was working than to force Herman to talk about biochemistry.[7]

At times, moreover, I was quite pleased with my current experiments on bacterial viruses. Within three months Ole and I had finished a set of experiments on the fate of a bacterial-virus particle when it multiplies inside a bacterium to form several hundred new virus particles. There were enough data for a respectable publication and, using ordinary standards, I knew I could stop work for the rest of the year without being judged unproductive. On the other hand, it was equally obvious that I had not done anything which was going to tell us what a gene was or how it reproduced. And unless I became a chemist, I could not see how I would.

VOL. 37, 1951   *BACTERIOLOGY: MAALØE AND WATSON*   507

THE TRANSFER OF RADIOACTIVE PHOSPHORUS FROM PARENTAL TO PROGENY PHAGE

BY O. MAALØE AND J. D. WATSON*

STATE SERUM INSTITUTE AND INSTITUTE OF CYTOPHYSIOLOGY, COPENHAGEN, DENMARK

Communicated by M. Delbrück, June 25, 1951

Introduction.—Reproduction is perhaps the most basic and characteristic feature of life. From the chemical point of view it is also the most obscure feature: atoms do not reproduce. When a living organism reproduces, there are now two atoms in the system for each one of the parent system. The additional atoms, of course, have not been "generated" by reproduction of the parent's atoms, but have been assimilated from the environment. Although the two progeny organisms may be biologically identical we should consider that their atoms can be classified into two classes: parental atoms and assimilated atoms. How are these atoms distributed between the two progeny organisms? Is one of the progeny all parental, the other all assimilated, or each half and half? Or perhaps both assimilated and the parental atoms dissimilated and passed into the environment? Are there specific macromolecular structures (genes?) that are preserved and passed on intact to the progeny? To answer questions of this kind we must be able to distinguish between parental and assimilated atoms and, in principle, this can be accomplished by the use of tracers.

Maaløe and Watson's paper in the Proceedings of the National Academy of Sciences *37: 507–513.*

[8] The Stazione Zoologica Napoli was founded by Anton Dohrn, a student of Ernst Haeckel, in 1872. It provided a center for researchers from around the world who studied embryology as a means of exploring evolutionary relationships. Among those who rented "tables" (laboratory benches) for summer seasons were T. H. Morgan, Hans Driesch, E. B. Wilson, and R. G. Harrison.

Stazione Zoologica Napoli.

I thus welcomed Herman's suggestion that I go that spring to the Zoological Station at Naples, where he had decided to spend the months of April and May.[8] A trip to Naples made great sense. There was no point in doing nothing in Copenhagen, where spring does not exist. On the other hand, the sun of Naples might be conducive to learning something about the biochemistry of the embryonic development of marine animals. It might also be a place where I could quietly read genetics. And when I was tired of it, I might conceivably pick up a biochemistry text. Without any hesitation I wrote to the States requesting permission to accompany Herman to Naples. A cheerful affirmative letter wishing me a pleasant journey came by return post from Washington. Moreover, it enclosed a $200 check for travel expenses. It made me feel slightly dishonest as I set off for the sun.

Chapter 4

[1] John Turton Randall was at the University of Birmingham when Wilkins joined him in 1938 as a Ph.D. student. During the war, Randall together with H. A. H. (Harry) Boot developed the cavity magnetron, an essential component of radar. After the war, Randall moved to St. Andrews University in 1944 and in 1946 to the Wheatstone Chair of Physics at King's College, London. After the war, Wilkins joined Randall at St. Andrews and went with him to King's. Randall was a highly accomplished administrator, Wilkins describing him as having "formidable entrepreneurial skills" which he used, with great success, to promote the activities of his scientists, whether or not their research was in his particular areas of interest. He was so successful that the Principal of King's College once complained to the MRC that Randall was receiving too much money and that the successes of his Biophysics Unit were distorting the College's overall research program.

The MRC Biophysics Unit was part of the Department of Physics, with Randall directing both. A large German bomb had destroyed the physics laboratories under the courtyard of King's College. These laboratories were rebuilt and opened in 1952.

Maurice Wilkins also had not come to Naples for serious science. The trip from London was an unexpected gift from his boss, Professor J. T. Randall. Originally Randall had been scheduled to come to the meeting on macromolecules and give a paper about the work going on in his new biophysics lab. Finding himself overcommitted, he had decided to send Maurice instead. If no one went, it would look bad for his King's College lab. Lots of scarce Treasury money had to be committed to set up his biophysics show, and suspicions existed that this was money down the drain.[1]

Randall at the annual department cricket match, 1950s.

No one was expected to prepare an elaborate talk for Italian meetings like this one. Such gatherings routinely brought together a small number of invited guests who did not understand Italian and a large number of Italians, almost none of whom under-

Workmen excavating the bomb crater in the quadrangle of King's College, London.

"Most of my time I spend walking the streets..." of Naples.

[2] Watson's pessimism was justified. Talks included "The rheology of the cross striated muscle and its minute structural interpretation" and "The cell wall–cytoplasm relationship in *Valonia*."

stood rapidly spoken English, the only language common to the visitors. The high point of each meeting was the day-long excursion to some scenic house or temple. Thus there was seldom chance for anything but banal remarks.

By the time Maurice arrived I was noticeably restless and impatient to return north. Herman had completely misled me. For the first six weeks in Naples I was constantly cold. The official temperature is often much less relevant than the absence of central heating. Neither the Zoological Station nor my decaying room atop a six-story nineteenth-century house had any heat. If I had had even the slightest interest in marine animals, I would have done experiments. Moving about doing experiments is much warmer than sitting in the library with one's feet on a table. At times I stood about nervously while Herman went through the motions of a biochemist, and on several days I even understood what he said. It made no difference, however, whether or not I followed the argument. Genes were never at the center, or even at the periphery, of his thoughts.

Most of my time I spent walking the streets or reading journal articles from the early days of genetics. Sometimes I daydreamed about discovering the secret of the gene, but not once did I have the faintest trace of a respectable idea. It was thus difficult to avoid the disquieting thought that I was not accomplishing anything. Knowing that I had not come to Naples for work did not make me feel better.

I retained a slight hope that I might profit from the meeting on the structures of biological macromolecules. Though I knew nothing about the X-ray diffraction techniques that dominated structural analysis, I was optimistic that the spoken arguments would be more comprehensible than the journal articles, which passed over my head. I was specially interested to hear the talk on nucleic acids to be given by Randall. At that time almost nothing was published about the possible three-dimensional configurations of a nucleic acid molecule. Conceivably this fact affected my casual pursuit of chemistry. For why should I get excited learning boring chemical facts as long as the chemists never provided anything incisive about the nucleic acids?

The odds, however, were against any real revelation then.[2] Much of the talk about the three-dimensional structure of proteins and nucleic acids was hot air. Though this work had been going on for over fifteen years, most if not all of the facts were soft. Ideas put forward with conviction were likely to be the products of wild crystallographers who delighted in being in a field where their ideas could not be easily disproved. Thus, although virtually all biochemists, including Herman, were unable to understand the arguments of the X-ray people, there was little uneasiness. It made no sense to learn complicated mathematical methods in order to follow baloney. As a result, none of my teachers had ever considered the possibility that I might do post-doctoral research with an X-ray crystallographer.

Maurice, however, did not disappoint me. The fact that he was a substitute for Randall made no difference: I had not known about either. His talk was far from vacuous and stood out sharply from the rest, several of which bore no connection to the purpose of the meeting. Fortunately these were in Italian, and so the obvious boredom of the foreign guests did not need to be construed as impoliteness. Several other speakers were continental biologists, at that time guests at the Zoological Station, who only briefly alluded to macromolecular structure. In contrast, Maurice's X-ray diffraction picture of DNA was to the point. It was flicked on the screen near the end of his talk. Maurice's dry English tone did not permit enthusiasm as he stated that the picture showed much more detail than previous pictures and could, in fact, be considered as arising from a crystalline substance. And when the structure of DNA was known, we might be in a better position to understand how genes work.[3]

Suddenly I was excited about chemistry. Before Maurice's talk I had worried about the possibility that the gene might be fantastically irregular.

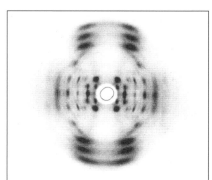

[3] *X-ray diffraction photograph of DNA taken by Wilkins and Raymond Gosling in 1950. This was the photograph which Wilkins showed in Naples.*

Gosling recalled (2012) his first sight of this image: "When…I first saw all those discrete diffraction spots emerging on the film in the developing dish was a truly eureka moment. Maurice and I drank several glasses of his sherry, kept in the bottom of one of his filing cabinets for V.I.P. visitors! We realized that if DNA was the gene material then we had just shown that genes could crystallize!"

(Left) *Gosling wound strands of DNA around a bent paper clip and took the diffraction picture shown above using the RayMax sealed X-ray tube in the Chemistry Department.* (Right) *X-ray tube used by Wilkins to take diffraction photographs.*

The temples at Paestum were built between 530 and 460 BC. Two temples were dedicated to the Goddess Hera and the third temple to the Goddess Athena.

Now, however, I knew that genes could crystallize; hence they must have a regular structure that could be solved in a straightforward fashion. Immediately I began to wonder whether it would be possible for me to join Wilkins in working on DNA. After the lecture I tried to seek him out. Perhaps he already knew more than his talk had indicated—often if a scientist is not absolutely sure he is correct, he is hesitant to speak in public. But there was no opportunity to talk to him; Maurice had vanished.

Not until the next day, when all the participants took an excursion to the Greek temples at Paestum, did I get an opportunity to introduce myself. While waiting for the bus I started a conversation and explained how interested I was in DNA. But before I could pump Maurice we had to board, and I joined my sister, Elizabeth, who had just come in from the States. At the temples we all scattered, and before I could corner Maurice again I realized that I might have had a tremendous stroke of good luck. Maurice had noticed that my sister was very pretty, and soon they were eating lunch together. I was immensely pleased. For years I had sullenly watched Elizabeth being pursued by a series of dull nitwits. Suddenly the possibility opened up that her way of life could be changed. No longer did I have to face the certainty that she would end up with a mental defective. Furthermore, if Maurice really liked my sister, it was inevitable that I would become closely associated

Elizabeth ("Betty") Watson (center) crossing the Atlantic, 1951.

with his X-ray work on DNA. The fact that Maurice excused himself to go and sit alone did not upset me. He obviously had good manners and assumed that I wished to converse with Elizabeth.

As soon as we reached Naples, however, my daydreams of glory by association ended. Maurice moved off to his hotel with only a casual nod. Neither the beauty of my sister nor my intense interest in the DNA structure had snared him. Our futures did not seem to be in London. Thus I set off to Copenhagen and the prospect of more biochemistry to avoid.[4]

[4] Watson wasn't the only one who felt his future did not lie in London. Upon returning from the Naples meeting, Wilkins told Gosling about Watson (describing him as a gangly young American). He instructed Gosling that if Watson showed up at King's, Gosling was to say that Wilkins had "left the country."

Kalckar's biochemistry laboratory in Copenhagen. Front row from left to right: Herman Kalckar, Audrey Jarnum, Jytte Heisel, Eugene Goldwasser, Walter McNutt, E. Hoff-Jorgensen. Back row: Gunther Stent, Niels Ole Kjeldgaard, Hans Klenow, Jim Watson, Vincent Price.

Chapter 5

I proceeded to forget Maurice, but not his DNA photograph. A potential key to the secret of life was impossible to push out of my mind. The fact that I was unable to interpret it did not bother me. It was certainly better to imagine myself becoming famous than maturing into a stifled academic who had never risked a thought. I was also encouraged by the very exciting rumor that Linus Pauling had partly solved the structure of proteins. The news hit me in Geneva, where I had stopped for several days to talk with the Swiss phage worker Jean Weigle, who was just back from a winter of work at Cal Tech. Before leaving, Jean had gone to the lecture where Linus had made the announcement.

Pauling's talk was made with his usual dramatic flair. The words came out as if he had been in show business all his life. A curtain kept his model hidden until near the end of his lecture, when he proudly unveiled his latest creation. Then, with his eyes twinkling, Linus explained the specific characteristics that made his model—the α-helix—uniquely beautiful.[1] This show, like all of his dazzling performances, delighted the younger students in attendance. There

Jean Weigle, 1951.

[1] In 1948, Pauling was a visiting professor at Oxford when he fell ill and, to while away the time as he lay in bed, decided to solve the structure of α-keratin. Using known chemical principles about bond lengths and angles, he drew a polypeptide chain on a piece of paper and then folded the paper to bring the appropriate groups in line to form hydrogen bonds. However, he did not publish a structure until 1951, by which time the α-helix model had been refined by the accurate determinations of the structures of amino acids and small peptides carried out by Pauling's group.

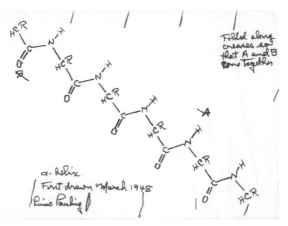

Pauling's recreation of the α-helix.

was no one like Linus in all the world. The combination of his prodigious mind and his infectious grin was unbeatable. Several fellow professors, however, watched this performance with mixed feelings. Seeing Linus jumping up and down on the demonstration table and moving his arms like a magician about to pull a rabbit out of his shoe made them feel inadequate. If only he had shown a little humility, it would have been so much easier to take! Even if he were to say nonsense, his mesmerized students would never know because of his unquenchable self-confidence. A number of his colleagues quietly waited for the day when he would fall flat on his face by botching something important.

Linus Pauling with his atomic models.

But Jean could not then tell me whether Linus' α-helix was right. He was not an X-ray crystallographer and could not judge the model professionally. Several of his younger friends, however, trained in structural chemistry, thought the α-helix looked very pretty. The best guess of Jean's acquaintances, therefore, was that Linus was right. If so, he had again accomplished a feat of extraordinary significance. He would be the first person to propose something solidly correct about the structure of a biologically important macromolecule. Conceivably, in doing so, he might have come up with a sensational new method which could be extended to the nucleic acids. Jean, however, did not remember any special tricks. The most he could tell me was that a description of the α-helix would soon be published.

By the time I was back in Copenhagen, the journal containing Linus' article had arrived from the States. I quickly read it and immediately reread it. Most of the language was above me, and so I could only get a general impression of his argument. I had no way of judging whether it made sense. The only thing I was sure of was that it was written with style. A few days later the next issue of the journal arrived, this time containing seven more Pauling articles. Again the language was dazzling and full of rhetorical tricks. One arti-

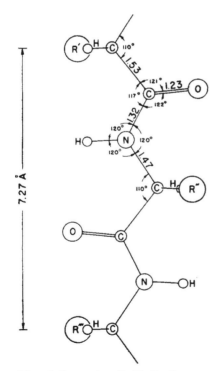

The α-helix was described by Pauling, Corey, and Branson in a paper published in the April 15, 1951 issue of Proceedings of the National Academy of Sciences.

CONTENTS

The table of contents page from the May 1951 issue of PNAS, *showing the seven papers from Pauling and Corey.*

cle started with the phrase, "Collagen is a very interesting protein." It inspired me to compose opening lines of the paper I would write about DNA, if I solved its structure. A sentence like "Genes are interesting to geneticists" would distinguish my way of thought from Pauling's.[2]

So I began worrying about where I could learn how to solve X-ray diffraction pictures. Cal Tech was not the place—Linus was too great a man to waste his time teaching a mathematically deficient biologist. Neither did I wish to be

[2] This influence of Pauling's style on Watson can perhaps be detected in the opening sentence of Watson and Crick's second paper on DNA, published in *Nature* in May 1953: "The importance of DNA within living cells is undisputed."

[3] In a letter dated August 9, 1951, Luria told Watson that he had spoken to Kendrew, who was visiting Indiana, and that Kendrew "...is quite anxious to have somebody like you and has space and money." Kendrew would let Perutz know that Watson might be in touch. Luria thought that Kendrew and Perutz were "sounder" than either Astbury (at Leeds) or Bernal (at Birkbeck College, London). What is more, "...chemistry & physics at Cambridge are good, and geneticists quite adequate—Roy Markham is a good man too..."

[4] As Watson wrote to his sister on July 14, 1951: "I am more certain about going to Cambridge next year." Watson's decision to complete his move to Cambridge before clearing it with the Merck Fellowship Board at the NRC was to lead to complications and frustration during his first six months in Cambridge.

further put off by Wilkins. This left Cambridge, England, where I knew that someone named Max Perutz was interested in the structure of the large biological molecules, in particular, the protein hemoglobin. I thus wrote to Luria about my newly found passion, asking whether he knew how to arrange my acceptance into the Cambridge lab. Unexpectedly, this was no problem at all. Soon after receiving my letter, Luria went to a small meeting at Ann Arbor, where he met Perutz' coworker, John Kendrew, then on an extended trip to the States. Most fortunately, Kendrew made a favorable impression on Luria; like Kalckar, he was civilized and in addition supported the Labor Party. Furthermore, the Cambridge

John Kendrew.

lab was understaffed and Kendrew was looking for someone to join him in his study of the protein myoglobin. Luria assured him that I would fit the bill and immediately wrote me the good news.[3]

It was then early August, just a month before my original fellowship would expire. This meant that I could not long delay writing to Washington about my change of plans. I decided to wait until I was admitted officially into the Cambridge lab. There was always the possibility that something would go wrong. It seemed prudent to put off the awkward letter until I could talk personally with Perutz. Then I could state in much greater detail what I might hope to accomplish in England. I did not, however, leave at once.[4] Again I was back in the lab, and the experiments I was doing were fun, in a second-class fashion. Even more important, I did not want to be away during the forthcoming International Poliomyelitis Conference, which was to bring several phage workers to Copenhagen. Max Delbrück was in the expected group, and since he was a professor at Cal Tech he might have further news about Pauling's latest trick.[5]

Delbrück, however, did not enlighten me further. The α-helix, even if correct, had not provided any biological insights; he seemed bored speaking about it. Even my information that a pretty X-ray photograph of DNA existed elicited no real response. But I had no opportunity to be depressed by Delbrück's characteristic bluntness, for the poliomyelitis congress was an unparalleled suc-

[5] A party at Niels (Taj) Jerne's house in September 1951 at the time of the International Poliomyelitis Conference in Copenhagen. Watson sits on the floor beside Florence Goldwasser. On his left is Niels Jerne holding a glass of beer. On the sofa behind is Ole Maaløe with his arm around Elizabeth Watson. We know from another photo taken within moments of this one but from another angle that Herman Kalckar and Barbara Wright are just out of shot on the left. Niels Jerne became an eminent immunologist, winning the Nobel Prize for Medicine in 1984.

cess. From the moment the several hundred delegates arrived, a profusion of free champagne, partly provided by American dollars, was available to loosen international barriers. Each night for a week there were receptions, dinners, and midnight trips to waterfront bars. It was my first experience with the high life, associated in my mind with decaying European aristocracy. An important truth was slowly entering my head: a scientist's life might be interesting socially as well as intellectually. I went off to England in excellent spirits.

A bar in the Tivoli Gardens, Copenhagen, 1952.

Chapter 6

Max Perutz was in his office when I showed up just after lunch. John Kendrew was still in the States, but my arrival was not unexpected. A brief letter from John said that an American biologist might work with him during the following year. I explained that I was ignorant of how X rays diffract, but Max immediately put me at ease. I was assured that no high-powered mathematics would be required: both he and John had studied chemistry as undergraduates. All I need do was read a crystallographic text; this would enable me to understand enough theory to begin to take X-ray photographs. As an example, Max told me about his simple idea for testing Pauling's α-helix. Only a day had been required to get the crucial photograph confirming Pauling's prediction. I did not follow Max at all. I was even ignorant of Bragg's Law, the most basic of all crystallographic ideas.[1]

We then went for a walk to look over possible digs for the coming year. When Max realized that I had come directly to the lab from the station and had not yet seen any of the colleges, he altered our course to take me through King's, along the backs, and through to the Great Court of Trinity. I had never seen such beautiful buildings in all my life, and any hesitation I might have

[1] Perutz had realized that if the α-helix was correct, it should produce a characteristic reflection at 1.5 Å arising from the amino acid repeat along the polypeptide chain. Perutz set up a camera with a suitable cylindrical film, and, using a horse hair as the sample, found the reflection confirming Pauling's model. He reported this in *Nature* on June 30, 1951.

View of King's College Chapel from the backs.

The Great Court of Trinity College.

[2] Watson's first impressions of Cambridge are made vivid in a letter written in September 1951, on his way back to Copenhagen following his visit to Perutz. Writing to his sister while standing up in a post office in London, Watson is clearly entranced by Cambridge: "The colleges at Cambridge are most impressive—In a way the most beautiful city I've ever been in...I believe I will like England a great deal."

Clare College.

had about leaving my safe life as a biologist vanished. Thus I was only nominally depressed when I peered inside several damp houses known to contain student rooms. I knew from the novels of Dickens that I would not suffer a fate the English denied themselves. In fact, I thought myself very lucky when I found a room in a two-story house on Jesus Green, a superb location less than ten minutes' walk from the lab.[2]

The following morning I went back to the Cavendish, since Max wanted me to meet Sir Lawrence Bragg. When Max telephoned upstairs that I was here, Sir Lawrence came down from his office, let me say a few words, and then retired for a private conversation with Max. A few minutes later they emerged to allow Bragg to give me his formal permission to work under his direction. The performance was uncompromisingly British, and I quietly concluded that the white-mustached figure of Bragg now spent most of its days sitting in London clubs like the Athenaeum.[3]

The thought never occurred to me then that later on I would have contact with this apparent curiosity out of the past. Despite his indisputable

[3] Watson's description of Lawrence Bragg in a letter to his sister is rather more expressive than in *The Double Helix*: "Bragg is [a] short, moderately plump person who sort of reminds you of the Colonel Blimp type." (September 1951)

The membership of the Athenaeum, a gentlemen's club founded in London in 1824, was composed of professionals, notably scientists, engineers, and physicians, but extended to clergy, writers, artists, and lawyers; both Charles Darwin and Charles Dickens were members. Perutz described how the future of his and Kendrew's work was decided at the Athenaeum:

"In traditional fashion, Bragg met Sir Edward Mellanby, the Secretary of the MRC, for luncheon at the Athenaeum Club. Bragg explained that Kendrew and I were out on a treasure hunt with only the remotest chance of success but that, if we did succeed, our results would provide an insight into the workings of life on the molecular scale. Even then it might take a very long time before they would bring any direct benefit to medicine. Mellanby took the risk."

Coffee room, Athenaeum, London, 1950s.

reputation, Bragg had worked out his Law just before World War I, so I assumed he must be in effective retirement and would never care about genes. I politely thanked Sir Lawrence for accepting me and told Max I would be back in three weeks for the start of the Michaelmas term. I then returned to Copenhagen to collect my few clothes and to tell Herman about my good luck in being able to become a crystallographer.

Herman was splendidly cooperative. A letter was dispatched telling the Fellowship Office in Washington that he enthusiastically endorsed my change in plans. At the same time I wrote a letter to Washington, breaking the news that my current experiments on the biochemistry of virus reproduction were at best interesting in a nonprofound way. I wanted to give up conventional biochemistry, which I believed incapable of telling us how genes work. Instead I told them that I now knew that X-ray crystallography was the key to genetics. I requested the approval of my plans to transfer to Cambridge so that I might work at Perutz' lab and learn how to do crystallographic research.[4]

I saw no point in remaining in Copenhagen until permission came. It would have been absurd to stay there wasting my time. The week before, Maaløe had departed for a year at Cal Tech, and my interest in Herman's type of biochemistry remained zero. Leaving Copenhagen was of course illegal in the formal sense. On the other hand, my request could not be refused. Everyone knew of Herman's unsettled state, and the Washington office must have been wondering how long I would care to remain in Copenhagen. Writing directly about Herman's absence from his lab would have been not only ungentlemanly but unnecessary.

[4] Kalckar's enthusiastic letter to Dr. C. J. Lapp of the Merck Fellowship Board of the NRC is dated October 5, 1951. "Since Dr. Watson has shown himself in possession of a fine judgment in selecting topics for his research, I feel certain that his decision to study under the guidance of Dr. M. F. Perutz at Cambridge University merits full support." Watson wrote his letter, too, the consequences of which were played out in the coming months, as alluded to in the text, and described in detail in Appendix 3.

```
    I should like to write in behalf of Dr. Watson and his
decision to study physics of high-molecular compounds.  Dr.
Watson and I have discussed various alternatives for some
time and I have contributed towards encouraging him to use
the second year of his fellowship to study at another labo-
ratory.
```

An extract from Kalckar's letter to Lapp at the NRC, October 5, 1951.

Naturally I was not at all prepared to receive a letter refusing permission. Ten days after my return to Cambridge, Herman forwarded the depressing news, which had been sent to my Copenhagen address. The Fellowship Board would not approve my transfer to a lab from which I was totally unprepared to profit. I was told to reconsider my plans, since I was unqualified to do crystallographic work. The Fellowship Board would, however, look favorably on a proposal that I transfer to the cell-physiology laboratory of Caspersson in Stockholm.

The source of the trouble was all too apparent. The head of the Fellowship Board no longer was Hans Clarke, a kindly biochemist friend of Herman's, then about to retire from Columbia. My letter had gone instead to a new chairman, who took a more active interest in directing young people. He was put out that I had overstepped myself in denying that I would profit from biochemistry. I wrote to Luria to save me. He and the new man were casual acquaintances, and so when my decision was set in proper perspective, he might reverse his decision.[5]

At first there were hints that Luria's interjection might cause a change back to reason. I was cheered up when a letter arrived from Luria that the situation might be smoothed over if we appeared to eat crow. I was to write Washington that a major inducement in my wanting to be in Cambridge was the presence of Roy Markham, an English biochemist who worked with plant viruses. Markham took the news quite casually when I walked into his office and told him that he might acquire a model student who would never bother him by cluttering up his lab with experimental apparatus. He regarded the scheme as a perfect example of the inability of Americans to know how to behave. Nonetheless, he promised to go along with this nonsense.[6]

Armed with the assurance that Markham would not squeal, I humbly wrote a long letter to Washington, outlining how I might profit from being in the joint presence of Perutz and Markham. At the end of the letter I thought it honest to break the news officially that I was in Cambridge and would remain there until a decision was made. The new man in Washington, however, did not play ball. The clue came when the return letter was addressed to Herman's

[5] Watson wrote to his sister on October 16, 1951: "They cannot see why I want to leave Copenhagen and so do not approve of my Cambridge idea...I have written Luria to handle the matter for me...As he wanted me to work with Perutz, I know he will fight for me. I do not intend to worry about the matter." In a later letter to his sister (November 28, 1951) he identifies the "new chairman" as Paul Weiss, an Austrian cell biologist at the University of Chicago, and remarks that Weiss "...has taken great offense" at Watson's machinations.

Roy Markham at "Viruses," the 1953 Cold Spring Harbor Symposium on Quantitative Biology.

[6] The letter from Luria laid out what needed to be done to placate Weiss and the NRC Merck Fellowship Board. Luria had not been impressed by Watson's communications with the NRC: "you goddam bastard, you wrote the silliest letter to the Committee about it!" Alluding to Watson's poor handwriting, Luria ended his letter with a heartfelt postscript: "From now on, I won't read any more letters from you unless typed. Understand?"

Luria's letter to Watson, October 20, 1951.

lab. The Fellowship Board was considering my case. I would be informed when a decision had been made. Thus it did not seem prudent to cash my checks, which were still sent to Copenhagen at the beginning of each month.

Fortunately, the possibility of my not being paid in the forthcoming year for working on DNA was only annoying and not fatal. The $3000 fellowship stipend that I had received for being in Copenhagen was three times that required to live like a well-off Danish student. Even if I had to cover my sister's recent purchase of two fashionable Paris suits, I would have $1000 left, enough for a year's stay in Cambridge. My landlady was also helpful. She threw me out after less than a month's residence. My main crime was not removing my shoes when I entered the house after 9:00 P.M., the hour at which her husband went to sleep. Also I occasionally forgot the injunction not to flush the toilet at similar hours and, even worse, I went out after 10:00 P.M. Nothing in Cambridge was then open, and my motives were suspect. John and Elizabeth Kendrew rescued me with the offer, at almost no rent, of a tiny room in their house on Tennis Court Road.[7] It was unbelievably damp and heated only by an aged electric heater. Nonetheless, I eagerly accepted the offer. Though it looked like an open invitation to tuberculosis, living with friends was infinitely preferable to any other digs I might find at this late moment. So without any reluctance I decided to stay at Tennis Court Road until my financial picture improved.[8]

[7] Watson had had doubts about his lodgings from the beginning. He wrote to his sister on October 9, 1951: "My lodgings (digs) are adequate though dull. I doubt that I will last long since the landlady appears to want a house of absolute silence and is quite unhappy about my desire to come in later than 10:30. I will have no difficulty in leaving when I find more attractive digs."

As he described in a later letter to his sister (January 28, 1952; misdated as 1951), things later improved: "My digs situation is somewhat unsettled but better. I'm staying with John Kendrew in a somewhat unfurnished house. The atmosphere is quite pleasant though and so I'm much better off than with my previous landlady."

[8] Despite the hardships over his housing, Watson quickly found much to enjoy in the cultural life of Cambridge, as reported in a letter to his sister of November 4, 1951: "Cambridge is as you can imagine superficially very quiet. All shops close early in the evening and bars such as Cafe Royal do not exist. Life is however, not very dull (or need not be)…On Thursday evening, I went to the local theater to see the Cocktail Party by T. S. Eliot. I must confess not being very fond of the play even though I enjoyed the evening. Eliot's attempts to solve the problems of life I do not find very satisfying—especially as I have no tendencies of being a saint. Perhaps the play is to be read instead of seen. From the viewpoint of movies, Cambridge is also quite good as some 8 cinemas exist here. At least 2 of them present foreign or 'intellectual' films and so I find that those are more than I have time to see. Tomorrow evening Myra Hess is to give a Sonata recital and for the coming week Tenessee Williams' Summer and Smoke is to be presented at the Arts Theater."

Chapter 7

[1] Watson enthused about Crick in a letter to Delbrück (December 9, 1951): "The most interesting member of the group is a research student named Francis Crick…He is no doubt the brightest person I have ever worked with and the nearest approach to Pauling I've ever seen…He never stops talking or thinking and since I spend much of my spare time in his house (he has a very charming French wife who is an excellent cook) I find myself in a state of suspended stimulation."

Crick also made clear the importance of the constant conversation with Watson: "If Jim had been killed by a tennis ball, I am reasonably sure I would not have solved the structure alone."

From my first day in the lab I knew I would not leave Cambridge for a long time. Departing would be idiocy, for I had immediately discovered the fun of talking to Francis Crick.[1] Finding someone in Max's lab who knew that DNA was more important than proteins was real luck. Moreover, it was a great relief for me not to spend full time learning X-ray analysis of proteins. Our lunch conversations quickly centered on how genes were put together. Within a few days after my arrival, we knew what to do: imitate Linus Pauling and beat him at his own game.

Pauling's success with the polypeptide chain had naturally suggested to Francis that the same tricks might also work for DNA. But as long as no one nearby thought DNA was at the heart of everything, the potential personal difficulties with the King's lab kept him from moving into action with DNA. Moreover, even though hemoglobin was not the center of the universe, Francis' previous two years at the Cavendish certainly had not been dull. More than enough protein problems kept popping up that required someone with a bent

Watson and Crick in a photograph with the Cavendish group, 1952.

Crick and Watson during a walk along the backs (1952). In the distance, King's College Chapel and to the left Clare College.

43

John Kendrew building a model of myoglo-bin, 1958.

[2] Kendrew began working on myoglo-bin, one-quarter the size of hemoglo-bin, in 1947. It was not until 1952, after testing myoglobin from seals, penguins, and the dugong, that he found that sperm whale myoglobin produced sufficiently good crystals for X-ray analysis. Kendrew published a low-resolution structure for myoglo-bin in 1958. He shared the 1962 Nobel Prize for Chemistry with Max Perutz.

toward theory. But now, with me around the lab always wanting to talk about genes, Francis no longer kept his thoughts about DNA in a back recess of his brain. Even so, he had no intention of abandoning his interest in the other labo-ratory problems. No one should mind if, by spending only a few hours a week thinking about DNA, he helped me solve a smashingly important problem.

As a consequence, John Kendrew soon realized that I was unlikely to help him solve the myoglobin structure. Since he was unable to grow large crystals of horse myoglobin, he initially hoped I might have a greener thumb. No ef-fort, however, was required to see that my laboratory manipulations were less skillful than those of a Swiss chemist. About a fortnight after my arrival in Cambridge, we went out to the local slaughterhouse to get a horse heart for a new myoglobin preparation. If we were lucky, the damage to the myoglobin molecules which prevented crystallization would be averted by immediately freezing the ex-racehorse's heart. But my subsequent attempts at crystallization were no more successful than John's. In a sense I was almost relieved. If they had succeeded, John might have put me onto taking X-ray photographs.[2]

No obstacle thus prevented me from talking at least several hours each day to Francis. Thinking all the time was too much even for Francis, and often when he was stumped by his equations he used to pump my reservoir of phage lore. At other moments Francis would endeavor to fill my brain with crystal-lographic facts, ordinarily available only through the painful reading of pro-fessional journals. Particularly important were the exact arguments needed to understand how Linus Pauling had discovered the α-helix.

I soon was taught that Pauling's accomplishment was a product of com-mon sense, not the result of complicated mathematical reasoning. Equations occasionally crept into his argument, but in most cases words would have sufficed. The key to Linus' success was his reliance on the simple laws of structural chemistry. The α-helix had not been found by only staring at X-ray pictures; the essential trick, instead, was to ask which atoms like to sit next to each other. In place of pencil and paper, the main working tools were a set of molecular models superficially resembling the toys of preschool children.

We could thus see no reason why we should not solve DNA in the same way. All we had to do was to construct a set of molecular models and begin to play—with luck, the structure would be a helix. Any other type of configuration would be much more complicated. Worrying about complications before ruling out the possibility that the answer was simple would have been damned foolishness. Pauling never got anywhere by seeking out messes.

From our first conversations we assumed that the DNA molecule contained a very large number of nucleotides linearly linked together in a regular

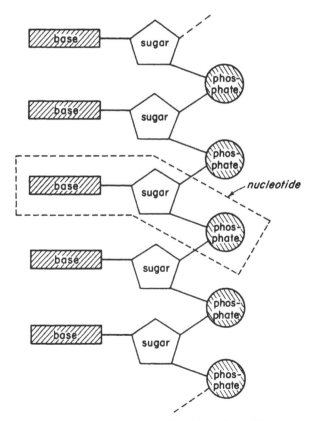

A short section of DNA as envisioned by Alexander Todd's research group in 1951. They thought that all the internucleotide links were phosphodiester bonds joining sugar carbon atom #5 to sugar carbon atom #3 of the adjacent nucleotide. As organic chemists they were concerned with how the atoms were linked together, leaving to crystallographers the problem of the 3-D arrangement of the atoms.

Alexander Todd and Linus Pauling in a punt on the River Cam in 1948.

[3] Alexander Todd was a Professor of Organic Chemistry at the University of Cambridge. He was notable for his synthesis of vitamins B1 and B12 for which he won the Nobel Prize for Chemistry in 1957. In the early 1950s, Todd was working on nucleotides and determining how they are linked together to make a polynucleotide chain.

way. Again our reasoning was partially based upon simplicity. Although organic chemists in Alexander Todd's nearby lab thought this the basic arrangement, they were still a long way from chemically establishing that all the internucleotide bonds were identical.[3] If this was not the case, however, we could not see how the DNA molecules packed together to form the crystalline aggregates studied by Maurice Wilkins and Rosalind Franklin. Thus, unless we found all future progress blocked, the best course was to regard the sugar-phosphate backbone as extremely regular and to search for a helical three-dimensional configuration in which all the backbone groups had identical chemical environments.

Immediately we could see that the solution to DNA might be more tricky than that of the α-helix. In the α-helix, a single polypeptide (a collection of amino acids) chain folds up into a helical arrangement held together by hydrogen bonds between groups on the same chain. Maurice had told Francis, however, that the diameter of the DNA molecule was thicker than would be the case if only one polynucleotide (a collection of nucleotides) chain were present. This made him think that the DNA molecule was a compound helix composed of several polynucleotide chains twisted about each other. If true, then before serious model building began, a decision would have to be made whether the chains would be held together by hydrogen bonds or by salt linkages involving the negatively charged phosphate groups.

A further complication arose from the fact that four types of nucleotides were found in DNA. In this sense, DNA was not a regular molecule but a highly irregular one. The four nucleotides were not, however, completely different, for each contained the same sugar and phosphate components. Their uniqueness lay in their nitrogenous bases, which were either a purine (ade-

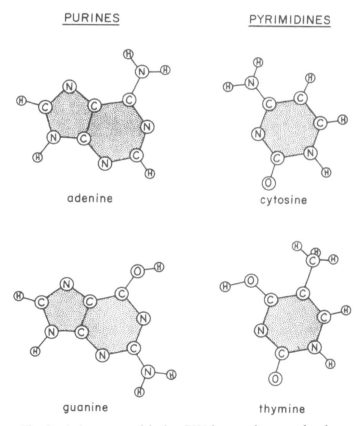

PURINES   PYRIMIDINES

adenine   cytosine

guanine   thymine

The chemical structures of the four DNA bases as they were often drawn about 1951. Because the electrons in the five- and six-membered rings are not localized, each base has a planar shape with a thickness of 3.4 Å.

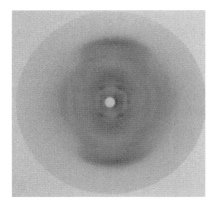 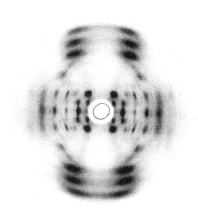

[4] *(Left)* *X-ray diffraction photograph of DNA taken by Florence Bell in 1938 (from her Ph.D. thesis) and* (right) *by Wilkins and Raymond Gosling in 1950 (see page 25). The improved resolution of Wilkins and Gosling's photograph is evident.*

W. T. Astbury, 1950s.

nine and guanine) or a pyrimidine (cytosine and thymine). But since the linkages between the nucleotides involved only the phosphate and sugar groups, our assumption that the same type of chemical bond linked all the nucleotides together was not affected. So in building models we would postulate that the sugar-phosphate backbone was very regular, and the order of bases of necessity very irregular. If the base sequences were always the same, all DNA molecules would be identical and there would not exist the variability that must distinguish one gene from another.

Though Pauling had got the α-helix almost without the X-ray evidence, he knew of its existence and to a certain extent had taken it into account. Given the X-ray data, a large variety of possible three-dimensional configurations for the polypeptide chain were quickly discarded. The exact X-ray data should help us go ahead much faster with the more subtly constructed DNA molecule. Mere inspection of the DNA X-ray picture should prevent a number of false starts. Fortunately, there already existed one half good photograph in the published literature.[4] It was taken five years previously by the English crystallographer W. T. Astbury, and could be used to start us off.[5] Yet possession of Maurice's much better crystalline photographs might save us from six months' to a year's work. The painful fact that the pictures belonged to Maurice could not be avoided.

Astbury and Florence Bell, 1939.

[5] Astbury worked with Sir William Henry Bragg at the Royal Institution in the 1920s. He moved to Leeds in 1928 and carried out pioneering structural studies of natural substances such as wool, hair, and porcupine quills.

Florence Bell joined Astbury in 1937, having carried out research in the Cavendish and the Department of Physics at Manchester. She was awarded a Ph.D. in 1939 for her studies of DNA.

There was nothing else to do but talk to him. To our surprise, Francis had no problem in persuading Maurice to come up to Cambridge for a weekend. And there was no need to force Maurice to the conclusion that the structure was a helix. Not only was it the obvious guess, but Maurice already had been talking in terms of helices at a summer meeting in Cambridge. About six weeks before I first arrived there, he had shown X-ray diffraction pictures of DNA which revealed a marked absence of reflections on the meridian. This was a feature that his colleague, the theoretician Alex Stokes, had told him was compatible with a helix. Given this conclusion, Maurice suspected that three polynucleotide chains were used to construct the helix.[6]

He did not, however, share our belief that Pauling's model-building game would quickly solve the structure, at least not until further X-ray results were obtained. Most of our conversation, instead, centered on Rosy Franklin. More trouble than ever was coming from her. She was now insisting that not even Maurice himself should take any more X-ray photographs of DNA. In trying to come to terms with Rosy, Maurice made a very bad bargain. He had handed over to her all the good crystalline DNA used in his

[6] Watson wrote to his sister about Wilkins' visit. He reported that Wilkins "...seems pleasant and at times somewhat funny with his obvious meditation on every step he takes." (November 14, 1951)

Maurice Wilkins, 1950s.

Vials containing DNA supplied to Wilkins by Rudolph Signer in 1950.

[7] The "good crystalline DNA" had been prepared by the Swiss chemist Rudolph Signer. After Franklin took over these samples, Wilkins began working with DNA supplied by Erwin Chargaff and was dismayed to find that Chargaff's DNA did not produce crystalline fibers.

original work and had agreed to confine his studies to other DNA, which he afterward found did not crystallize.[7]

The point had been reached where Rosy would not even tell Maurice her latest results. The soonest Maurice was likely to learn where things stood was three weeks hence, the middle of November. At that time Rosy was scheduled to give a seminar on her past six months' work. Naturally I was delighted when Maurice said I would be welcome at Rosy's talk. For the first time I had a real incentive to learn some crystallography: I did not want Rosy to speak over my head.

Chapter 8

Most unexpectedly, Francis' interest in DNA temporarily fell to almost zero less than a week later. The cause was his decision to accuse a colleague of ignoring his ideas. The accusation was leveled at none other than his Professor. It happened less than a month after my arrival, on a Saturday morning. The previous day Max Perutz had given Francis a new manuscript by Sir Lawrence and himself, dealing with the shape of the hemoglobin molecule. As he rapidly read its contents Francis became furious, for he noticed that part of the argument depended upon a theoretical idea he had propounded some nine months earlier. What was worse, Francis remembered having enthusiastically proclaimed it to everyone in the lab. Yet his contribution had not been acknowledged.[1] Almost at once, after dashing in to tell Max and John Kendrew about the outrage, he hurried to Bragg's office for an explanation, if not an apology. But by then Bragg was at home, and Francis had to wait until the following morning. Unfortunately, this delay did not make the confrontation any more successful.

Sir Lawrence Bragg sitting at his Cavendish desk.

[1] This research was published in three papers in 1952. The only acknowledgment of Crick's help came in one paper: "We wish to acknowledge the help of...Mr. F. H. C. Crick and Mr. H. E. Huxley who took some of the required photographs."

Sir Lawrence flatly denied prior knowledge of Francis' efforts and was thoroughly insulted by the implication that he had underhandedly used another scientist's ideas. On the other hand, Francis found it impossible to believe that Bragg could have been so dense as to have missed his oft-repeated idea, and he as much as told Bragg this. Further conversation became impossible, and in less than ten minutes Francis was out of the Professor's office.

The courtyard of the Eagle, Bene't Street, just one hundred yards from the Cavendish Laboratory, 1937.

For Bragg this meeting seemed the final straw in his relations with Crick. Several weeks earlier Bragg had come into the lab greatly excited about an idea that came to him the previous evening, one that he and Perutz subsequently incorporated in their paper. While he was explaining it to Perutz and Kendrew, Crick happened to join the group. To his considerable annoyance, Francis did not accept the idea immediately but instead stated that he would go away and check whether Bragg was right or wrong. At this stage Bragg had blown his top and, with his blood pressure all too high, returned home presumably to tell his wife about the latest antics of their problem child.

This most recent tussle was a disaster for Francis, and he showed his uneasiness when he came down to the lab. Bragg, in dismissing him from his room, had angrily told him that he would consider seriously whether he could continue to give Francis a place in the laboratory after his Ph.D. course was ended. Francis was obviously worried that he might soon have to find a new position. Our subsequent lunch at the Eagle, the pub at which he usually ate, was restrained and unpunctuated by the usual laughter.

His concern was not without reason. Although he knew he was bright and could produce novel ideas, he could claim no clear-cut intellectual

Crick relaxing on the roof of University College, London.

achievements, and he was still without his Ph.D. He came from a solid middle-class family and was sent to school at Mill Hill. Then he read physics at University College, London, and had commenced work on an advanced degree when the war broke out. Like almost all other English scientists, he joined the war effort and became part of the Admiralty's scientific establishment. There he worked with great vigor, and, although many resented his nonstop conversation, there was a war to win and he was quite helpful in producing ingenious magnetic mines.[2] When the war was over, however, some of his colleagues saw no sound reason to have him about forever, and for a period he was given to believe that he had no future in the scientific civil service.

Moreover, he had lost all desire to stay in physics and decided instead to try biology. With the help of the physiologist A. V. Hill, he obtained a small grant to come up to Cambridge in the fall of 1947.[3] At first he did true biology at the Strangeways Laboratory, but this was obviously trivial and two years later he moved over to the Cavendish, where he joined Perutz and Kendrew.[4,5] Here he again became excited about science and decided that

[2] The Physics Laboratory at University College, London, was closed for the duration of the war and relocated in shared accommodation in the Physics department at the University College of North Wales, now Bangor University. Crick was recruited to the Admiralty Research Laboratory and assigned to the Mine Design Department. He spent the war years very successfully designing firing mechanisms for mines.

[3] *A. V. Hill, muscle physiologist, Nobel Prize in Physiology or Medicine, 1922, who advised Crick to do research in biology.*

[4] Thomas Pigg Strangeways founded a research hospital in 1905 and constructed a laboratory in 1912. This was designed as a house so that in the event of financial difficulties, it could be converted to a family residence and sold.

Strangeways Laboratory.

[5] Crick studied the viscosity of cytoplasm in the Strangeways Laboratory of Arthur Hughes. Cells in tissue culture took up iron particles which were then moved through the cytoplasm using a magnet. He published two papers in *Experimental Cell Research,* one experimental with Hughes, the other theoretical by himself.

with the authors compliments *fhcc*

Watson

Reprinted from Experimental Cell Research, Vol. 1 No. 1, 1950.

THE PHYSICAL PROPERTIES OF CYTOPLASM

A STUDY BY MEANS OF THE MAGNETIC PARTICLE METHOD

Part I. Experimental

F. H. C. CRICK and A. F. W. HUGHES

Strangeways Research Laboratory, Cambridge, England

Received October 5, 1949

A. INTRODUCTION

THIS PAPER has two aims: first to describe our experimental work on the physical properties of the cytoplasm of chick fibroblasts in tissue culture, with a short discussion of its implications and secondly to present such details of the magnetic particle method as to enable other workers to use it.

A second part will be devoted to the theory underlying these methods.

Watson

Reprinted from Experimental Cell Research, Vol. 1, No. 4, 1950.

THE PHYSICAL PROPERTIES OF CYTOPLASM. A STUDY BY MEANS OF THE MAGNETIC PARTICLE METHOD. PART II. THEORETICAL TREATMENT

F. H. C. CRICK

Strangeways Research Laboratory, Cambridge

Received February 9, 1950

A. INTRODUCTION

IN Part I (1) a method was described for measuring some of the physical properties of the cytoplasm of chick cells in tissue culture by means of magnetic particles. The cells were allowed to phagocytose these particles, which were then acted on by magnetic fields, their movements being observed simultaneously under high magnification.

In this paper the theoretical basis for the experimental methods used has been set out. The results are mainly standard pieces of magnetism and hydrodynamics, but as they are scattered about in the literature it was thought worth while to bring them all together in one place.

Watson's copies of Crick's first papers in biology.

Biophysics Research Unit
UNIVERSITY COLLEGE LONDON
GOWER STREET, W.C.1.
EUSTON 4400

AVH/JVF   11th March, 1949.

Dear Crick,

        Thank you for letting me know about your
decision.   I expect you are quite right.   If the X-ray
diffraction studies of protein are what interest you most,
in spite of any deterrent I may have exerted, you can be
reasonably sure that your decision is the best one.

[6] A letter from A. V. Hill responding to the news that Crick is going to work on protein structure. Hill consoles himself that "...it will be a good thing to have someone mixed up with it who does know something about the properties of living material." Crick worked with Perutz, trying to see what could be learned of the structure of hemoglobin by X-ray analysis of crystals containing different amounts of water. His Ph.D. was titled "X-ray diffraction: polypeptides and proteins."

perhaps he should finally work for a Ph.D. He thus enrolled as a research student (of Caius College) with Max as his supervisor.[6] In a sense, this pursuit of the Ph.D. was a bore to a mind that worked too fast to be satisfied with the tedium involved in thesis research. On the other hand, his decision had yielded an unforeseen dividend: in this moment of crisis, he could hardly be dismissed before he got his degree.

Max and John quickly came to Francis' rescue and interceded with the Professor. John confirmed that Francis had previously written an account of the argument in question, and Bragg acknowledged that the same idea had occurred independently to both. Bragg by that time had calmed down, and any question of Crick's going was quietly shelved. Keeping him on was not easy on Bragg. One day, in a moment of despair, he revealed that Crick made his ears buzz. Moreover, he remained unconvinced that Crick was needed. Already for thirty-five years he had not stopped talking and almost nothing of fundamental value had emerged.[7]

[7] Bragg regarded Crick as a protégé of A. V. Hill and made his concerns about Crick known to Hill in a letter dated January 18, 1952: "I am worried about him [Crick]...My worry is that it is almost impossible to get him to settle down to any steady job and I doubt whether he has enough material for his Ph.D. which should be taken this year...I should like some help in deciding what line to take with him."

Chapter 9

A new opportunity to theorize soon brought Francis back to normal form. Several days after the fiasco with Bragg, the crystallographer V. Vand sent Max a letter containing a theory for the diffraction of X rays by helical molecules. Helices were then at the center of the lab's interest, largely because of Pauling's α-helix. Yet there was still lacking a general theory to test new models as well as to confirm the finer details of the α-helix. This is what Vand hoped his theory would do.

Francis quickly found a serious flaw in Vand's efforts, became excited about finding the right theory, and bounded upstairs to talk with Bill Cochran, a small, quiet Scot, then a lecturer in crystallography at the Cavendish. Bill was the cleverest of the younger Cambridge X-ray people, and even though he was not involved in work on the large biological macromolecules, he always provided the most

Bill Cochran.   *Vladimir Vand.*

astute sounding board for Francis' frequent ventures into theory. When Bill told Francis that an idea was unsound or would lead nowhere, Francis could be sure that professional jealousy was not involved. This time, however, Bill did not voice skepticism, since independently he had found faults in Vand's paper and had begun to wonder what the right answer was. For months both Max and Bragg had been after him to work out the helical theory, but he had not moved into action. Now, with the additional pressure from Francis, he too began seriously to ponder how the equations should be set up.[1]

[1] Vladimir Vand was a Czech crystallographer who later, at the University of Pennsylvania, studied samples of lunar materials.

Bill Cochran was reluctant to work out the helical theory because he "…was convinced that it was impossible to determine protein structure by X-ray methods."

Crick and Odile (née Speed) on their wedding day, August 13, 1949. Odile had been a Wren (phonetic pronunciation of Women's Royal Naval Services–WRNS) when Crick met her in 1945. They were married for 55 years.

The remainder of the morning Francis was silent and absorbed in mathematical equations. At lunch at the Eagle a bad headache came on, and he went home instead of returning to the lab. But sitting in front of his gas fire doing nothing bored him, and again he took up his equations. To his delight, he soon saw that he had the answer. Nonetheless, he stopped his work, for he and his wife, Odile, were invited to a wine tasting at Matthews', one of Cambridge's better wine merchants. For several days his morale had been buoyed by the request to sample the wines. It meant acceptance by a more fashionable and amusing part of Cambridge and allowed him to dismiss the fact that he was not appreciated by a variety of dull and pompous dons.

He and Odile were then living at the "Green Door," a tiny, inexpensive flat on top of a several hundred-year-old house just across Bridge Street from St. John's College. There were only two rooms of any size, a living room and a bedroom. All the others, including the kitchen, in which the bathtub was the largest and most conspicuous object, were almost nonexistent. But despite the cramp, its great charm, magnified by Odile's decorative sense, gave

Matthew & Son Ltd, "one of Cambridge's better wine merchants."

The "Green Door." The wooden edifice on the left conceals a staircase which led up to the Cricks' minuscule top floor flat.

it a cheerful, if not playful, spirit. Here I first sensed the vitality of English intellectual life, so completely absent during my initial days in my Victorian room several hundred yards away on Jesus Green.

They had then been married for three years. Francis' first marriage did not last long, and a son, Michael, was looked after by Francis' mother and aunt.[2] He had lived alone for several years until Odile, some five years his junior, came to Cambridge and hastened his revolt against the stodginess of the middle classes, which delight in unwicked amusements like sailing and tennis, habits particularly unsuited to the conversational life. Neither was politics or religion of any concern. The latter was clearly an error of past generations, which Francis saw no reason to perpetuate. But I am less certain about their complete lack of enthusiasm for political issues. Perhaps it was the war, whose grimness they now wished to forget. In any case, *The Times* was not present at breakfast, and more attention was given to *Vogue*, the only magazine to which they subscribed and about which Francis could converse at length.[3]

By then I was often going to the Green Door for dinner. Francis was al-

[2] Crick had married Doreen Dodd on February 18, 1940 and Michael was born in November of that year. For much of the war, Crick was stationed in Havant, away from his family. By January 1946, the marriage had ended.

[3] Crick wrote later:

"I formed the opinion that it [politics] was not a very interesting thing unless one was especially informed about what was going on. It was for this reason that we have never had a daily paper since then and I never read *The Times* at breakfast as you imply." (Letter to Watson, March 31, 1966 commenting on an early draft of the book.)

4 According to the Oxford English Dictionary, a popsy is "An endearing appellation for a young girl" and was used first in 1862, in the children's book, *Pippins & Pies*.

5 Odile described the effect a pretty woman could have on Crick:

"Her very presence brings out in Francis a show of verbal fireworks and the compulsion to make an impression takes over completely."

ways eager to continue our conversations, while I joyously seized every opportunity to escape from the miserable English food that periodically led me to worry about whether I might have an ulcer. Odile's French mother had imparted to her a thorough contempt for the unimaginative way in which most Englishmen eat and house themselves. Francis thus never had reason to envy those college fellows whose High Table food was undeniably better than their wives' drab mixtures of tasteless meat, boiled potatoes, colorless greens, and typical trifles. Instead, dinner was often gay, especially after the wine turned the conversation to the currently talked-about Cambridge popsies.[4]

There was no restraint in Francis' enthusiasms about young women—that is, as long as they showed some vitality and were distinctive in any way that permitted gossip and amusement. When young, he saw little of women and was only now discovering the sparkle they added to life. Odile did not mind this predilection, seeing that it went along with, and probably helped, his emancipation from the dullness of his Northampton upbringing.[5] They would talk at length about the somewhat artsy-craftsy world in which Odile moved and into which they were frequently invited. No choice event was kept out of our conversations, and he would show equal gusto in telling of his occasional mistakes. One occurred when there was a costume party and he went looking like a young G. B. Shaw in a full red beard. As soon as he entered he realized that it was a ghastly error, since not one of the young women enjoyed being tickled by the wet, scraggly hairs when he came within kissing distance.

But there were no young women at the wine tasting. To his and Odile's dismay, their companions were college dons contentedly talking about the burdensome administrative problems with which they were so sadly afflicted. They went home early and Francis, unexpectedly sober, thought more about his answer.

The next morning he arrived in the lab and told Max and John about his success. A few minutes later, Bill Cochran walked into his office, and Francis started to repeat the story. But before he could let loose his argument, Bill told him that he also thought he had succeeded. Hurriedly they went through their respective mathematics and discovered that Bill had used an elegant derivation compared to Francis' more laborious approach. Gleefully, however, they found that they had arrived at the same final answer. They then checked the α-helix by visual inspection with Max's X-ray diagrams. The agreement was so good that both Linus' model and their theory had to be correct.[6]

Within a few days a polished manuscript was ready and jubilantly dispatched to *Nature*. At the same time, a copy was sent to Pauling to appreciate. This event, his first unquestionable success, was a signal triumph for Francis. For once the absence of women had gone along with luck.[7]

[6] Cochran wrote later that it was not Perutz' photographs of hemogloblin which provided the experimental evidence. Rather, Bragg had given Cochran crystals of poly-methyl glutamate and it was photographs of these, taken by Cochran, which provided the experimental data for checking their calculations.

[7] Later, Crick, writing to Watson about his objections to *The Double Helix*, dismissed Watson's contention that the absence of women that evening was of any significance:

"As far as I recall I finished all the algebra before I went to the wine tasting and your wonderful generalisation about the absence of women bringing me luck I don't think has any foundation in fact at all." (Letter to Watson, March 31, 1966)

A detail from a 1949 photograph of the members of the Cavendish Laboratory. Front row, 2nd from right, Bill Cochran; middle row, extreme left, John Kendrew, 2nd from right, Francis Crick, 1st from right, Max Perutz; Back row, 2nd from right, Hugh Huxley.

The Cochran-Crick paper on the Fourier transform of α-helix published in Nature, *February 9, 1952.*

234 NATURE February 9, 1952 VOL. 169

LETTERS TO THE EDITORS

The Editors do not hold themselves responsible for opinions expressed by their correspondents. No notice is taken of anonymous communications

Evidence for the Pauling–Corey α-Helix in Synthetic Polypeptides

WE have calculated, in collaboration with Dr. V. Vand[1], the Fourier transform (or continuous structure factor) of an atom repeated at regular intervals on an infinite helix. The properties of the transform are such that it will usually be possible to predict the general character of X-ray scattering by any structure based on a regular succession of similar groups of atoms arranged in a helical manner. In particular, the type of X-ray diffraction picture given by the synthetic polypeptide poly-γ-methyl-L-glutamate, which has been prepared in a highly crystalline form by Dr. C. H. Bamford and his colleagues in the Research Laboratories, Courtaulds, Ltd., Maidenhead, is so readily explained on this basis as to leave little doubt that the Pauling-Corey α-helix[2], or some close approximation to it, exists in this polypeptide. Pauling and Corey[2] have already shown this correspondence in the equatorial plane; it is shown here that the correspondence extends over the whole of the diffraction pattern.

We quote here the value of the transform which applies when the axial distance between successive turns of the helix is P, the axial distance between the successive atoms lying on the helix is p, and the structure so formed is repeated exactly in an axial distance c. (For the latter condition to be possible, P/p must be expressible as the ratio of whole numbers.) In this case, the transform is restricted to planes in reciprocal space which are perpendicular to the axis of the helix, and occur at heights $\zeta = l/c$, where l is an integer. In crystallographic nomenclature, these are the layer lines corresponding to a unit cell of length c. On the lth such plane the transform has the value :

$$F\left(R,\psi,\frac{l}{c}\right) = f \sum_n J_n(2\pi Rr) \exp\left[in\left(\psi + \frac{\pi}{2}\right)\right]. \quad (1)$$

(R,ψ,ζ) are the cylindrical co-ordinates of a point in reciprocal space, f is the atomic scattering factor, and J_n is the Bessel function of order n; r is the radius of the helix on which the set of atoms lies, the axes in real space being chosen so that one atom lies at $(r,0,0)$. For a given value of l, the sum in equation (1) is to be taken over all integer values of n which are solutions of the equation,

$$\frac{n}{P} + \frac{m}{p} = \frac{l}{c}, \quad (2)$$

m being any integer[1].

Thus only certain Bessel functions contribute to a particular layer line. This is illustrated in the accompanying table for the case of poly-γ methyl-L-glutamate, for which Pauling and Corey[2] suggested $P = 5\cdot4$ A., $p = 1\cdot5$ A. and $c = 27$ A. The first column lists the number, l, of the layer line, while the second gives the orders (n) of the Bessel functions which contribute to it (for simplicity only the lowest two values of n are given for each layer line).

Now there is, of course, more than one set of atoms in the polypeptide, but for all of them, P, p and c are the same, although r is different. The basis of

Value of l for the layer line	Lowest two values of n allowed by theory		Observed average strength of layer line (ref. 4)
0	**0**	± 18	strong
1	− 7	+ 11	
2	+ 4	− 14	*weak
3	− 3	+ 15	very weak
4	+ 8	− 10	medium
5	+ 1	− 17	
6	− 6	+ 12	
7	+ 5	− 13	
8	− 2	+ 16	weak
9	± 9		weak
10	+ 2	− 16	weak
11	− 5	+ 13	
12	+ 6	− 12	
13	− 1	+ 17	very weak
14	− 8	+ 10	
15	+ 3	− 15	
16	− 4	+ 14	
17	+ 7	− 11	
18	**0**	+ 18	medium
19	− 7	+ 11	
20	+ 4	− 14	
21	− 3	+ 15	
22	+ 8	− 10	
23	+ 1	− 17	trace
24	− 6	+ 12	
25	+ 5	− 13	
26	− 2	+ 16	trace
27	± 9		
28	+ 2	− 16	trace

Layers not described are absent.

* (10$\bar{1}$2), the reflexion having the smallest value of R, is absent.

our prediction is that a reflexion will be absent if the contribution of all sets of atoms to it is very small, and that on the average it will be strong if all sets of atoms make a large contribution.

It is a property of Bessel functions of higher order, illustrated in the graph, that they remain very small until a certain value of $2\pi Rr$ is reached, and that this point recedes from the origin as the order increases. Now, whatever the precise form of the chain, the value of r for any atom cannot be greater than about 8 A. because of the packing of the chains. This is a limit to the value of $2\pi Rr$ within the part of the transform covered by the observed diffraction picture $(R < 0\cdot3$ A.$^{-1}$ for $l \neq 0)$. No set of atoms can make an appreciable contribution to the amplitude of a reflexion occurring on a layer line with which only high-order Bessel functions are associated, because $2\pi Rr$ comes within the very low part of the curve in the graph.

We should therefore predict that layer lines to which only high-order Bessel functions contribute would be weak or absent, and that those to which very low orders contribute would be strong.

These predictions are strikingly borne out by the experimental data[4] summarized in the last column of the table. The significant Bessel functions involved in the first twenty-eight layer lines are shown in the second column, and, as will be seen, only layer lines associated with a function of order **4** or less

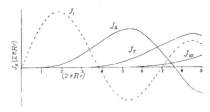

The march of higher-order Bessel functions (with J_1 added dashed)

Chapter 10

[1] The day before, Watson had written to his parents mentioning Franklin's seminar as one event among many filling his time: "On Saturday night, I went to a quite excellent party at Trinity College. On Sunday I recovered from the party. Yesterday evening I went to a sherry party at Professor Bragg's house. Later this evening I will attend a lecture on Human Genetics at King's. Tomorrow I will go into London to hear a lecture on Nucleic Acids at King's College, London. On Thursday, I [intend?] to hear two different lectures on Biochemistry. Then on Friday I will probably go to Oxford for a visit since others from the lab are going."

By mid-November, when Rosy's talk on DNA rolled about, I had learned enough crystallographic argument to follow much of her lecture. Most important, I knew what to focus attention upon. Six weeks of listening to Francis had made me realize that the crux of the matter was whether Rosy's new X-ray pictures would lend any support for a helical DNA structure. The really relevant experimental details were those which might provide clues in constructing molecular models. It took, however, only a few minutes of listening to Rosy to realize that her determined mind had set upon a different course of action.[1]

She spoke to an audience of about fifteen in a quick, nervous style that suited the unornamented old lecture hall in which we were seated. There was not a trace of warmth or frivolity in her words. And yet I could not regard her as totally uninteresting. Momentarily I wondered how she would

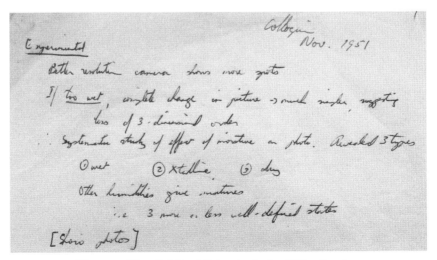

The start of Rosalind Franklin's notes for her 1951 colloquium.

Franklin in happier times with French colleagues. **Left,** *with Jacques Mering, late 1940s.* **Right,** *with Vittorio Luzzati, 1951.*

[2] Watson's notorious first impressions of Franklin are belied by other accounts of her. Certainly it is a happier figure we see in these two photos. Even during her time at King's others saw not only her combative and unhappy side, but a more sophisticated, well-dressed colleague. As Raymond Gosling, her Ph.D. student at King's, said in his 2010 interview on the BBC radio 4 Today program, "He [Watson] never saw her going out in the evening with the lead violinist of the symphony orchestra. She led a life, a social life, on a higher plane than the rest of us."

look if she took off her glasses and did something novel with her hair. Then, however, my main concern was her description of the crystalline X-ray diffraction pattern.[2]

The years of careful, unemotional crystallographic training had left their mark. She had not had the advantage of a rigid Cambridge education only to be so foolish as to misuse it. It was downright obvious to her that the only way to establish the DNA structure was by pure crystallographic approaches. As model building did not appeal to her, at no time did she mention Pauling's triumph over the α-helix. The idea of using tinker-toy-like models to solve biological structures was clearly a last resort. Of course Rosy knew of Linus' success but saw no obvious reason to ape his mannerisms. The measure of his past triumphs was sufficient reason in itself to act differently; only a genius of his stature could play like a ten-year-old boy and still get the right answer.

Rosy regarded her talk as a preliminary report which, by itself, would not test anything fundamental about DNA. Hard facts would come only when further data had been collected which could allow the crystallographic analyses to be carried to a more refined stage.[3] Her lack of immediate optimism was shared by the small group of lab people who came to the talk.

20.12.51

Swelling of fibres.

Fibres of "non-pulling" Signer DNA, swelling observed under microscope

(1) Small fibre ~ 0.6 scale divisions (i.e. 8 μ) at narrowest best oriented part enclosed in 92% humidity (sat. Na_2CO_3)

Rapid swelling to ~ 2x diameter – well-oriented part rather less

(2) Fibre of unequal quality, finest and best part ~ 8 μ,

Enclosed by water

Rapid swelling → ~ 2x diameter

then slow

after 2 hours, diameter increased ~ x 10.

length increase small (straight, ~~stretched~~ fibre, with ends fixed, → only slightly wavy)

Small drops of condensed water visible everywhere except immediately on either side of fibre (showing that even for 10-fold lateral swelling reduction of VP is appreciable)

11.12.51 → 28.12.51

2 attempts to photograph single fibres ~10 μ after baking out, one exposure 2 weeks, one 1 week

→ v. poor intensity, much impurity and only disordered DNA (used new "non-pulling" Signer specimen)

(28)

27.12.51 Single fibre (old Signer DNA) after 3 weeks over P_2O_5 (pulls v. well) 0.001" Al over ½ film, 3 films

Baked over-night at 75° (6 p.m. – 10 a.m.)

28.12.51 5 p.m. – camera, H_2, $NaClO_3$

6.00 tube on (new filament)

29.12.51 12 noon beam v. weak

31.12.51 10 a.m. beam invisible. Developed

→ wet photograph

∴ specimen baked out overnight before next photograph

(29) 67% RH 3 films

Poor resolution

(30) As (29), 68% RH 1 film

still poor resolution

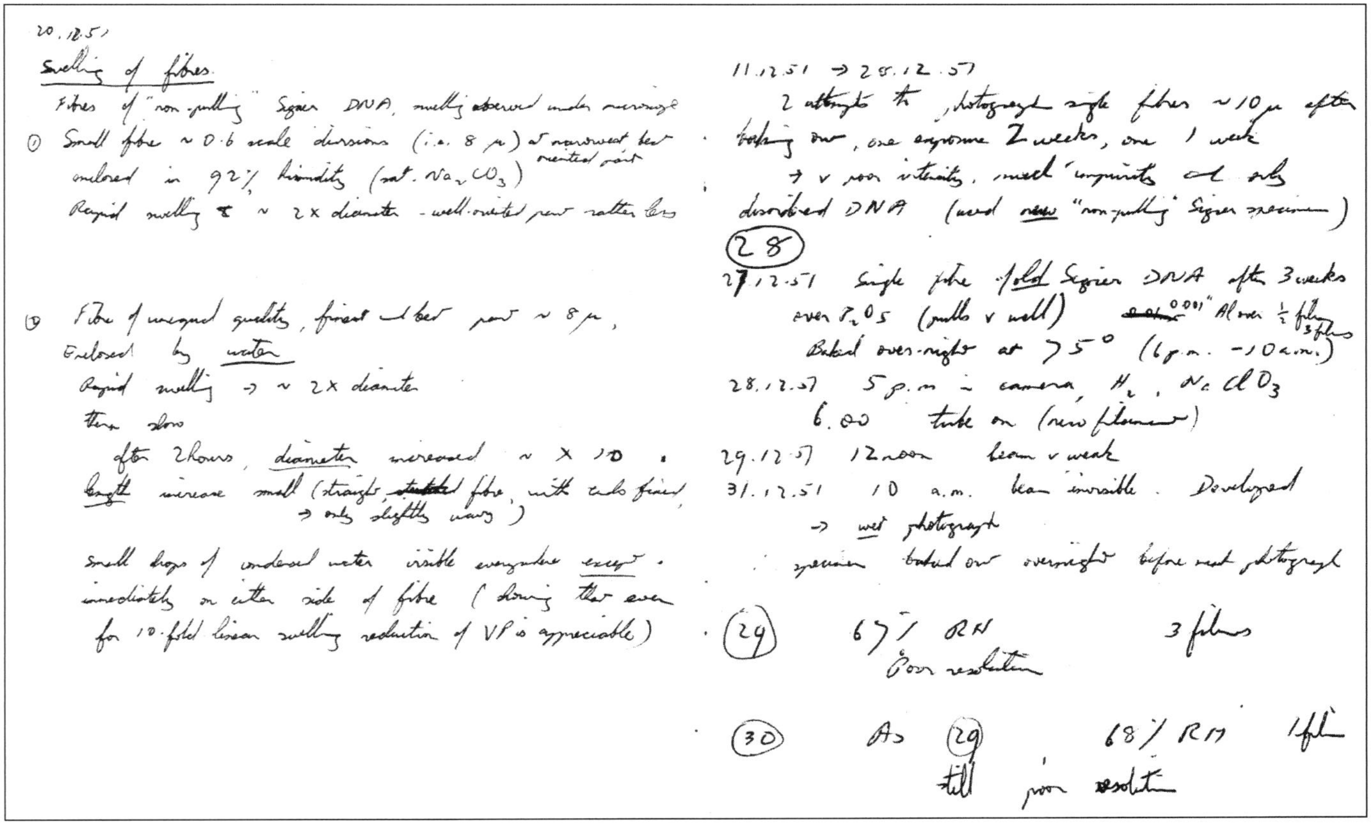

No one else brought up the desirability of using molecular models to help solve the structure. Maurice himself only asked several questions of a technical nature. The discussion then quickly stopped with the expressions on the listeners' faces indicating either that they had nothing to add or that, if they did wish to say something, it would be bad form since they had said it before. Maybe their reluctance to utter anything romantically optimistic, or even to mention models, was due to fear of a sharp retort from Rosy. Certainly a bad way to go out into the foulness of a heavy, foggy November

[3] **Shown above are pages from Franklin's laboratory notebook, describing experiments on the swelling of DNA, carried out in December 1951. Similar experiments, studying DNA in varying degrees of humidity, led to Franklin's discovery of the A ("dry") and B ("wet") forms of DNA, a distinction of great importance later in the story.**

A "heavy, foggy November night," London in the 1950s.

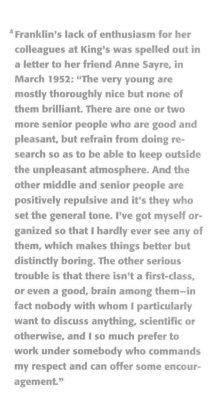

[4] Franklin's lack of enthusiasm for her colleagues at King's was spelled out in a letter to her friend Anne Sayre, in March 1952: "The very young are mostly thoroughly nice but none of them brilliant. There are one or two more senior people who are good and pleasant, but refrain from doing research so as to be able to keep outside the unpleasant atmosphere. And the other middle and senior people are positively repulsive and it's they who set the general tone. I've got myself organized so that I hardly ever see any of them, which makes things better but distinctly boring. The other serious trouble is that there isn't a first-class, or even a good, brain among them—in fact nobody with whom I particularly want to discuss anything, scientific or otherwise, and I so much prefer to work under somebody who commands my respect and can offer some encouragement."

night was to be told by a woman to refrain from venturing an opinion about a subject for which you were not trained. It was a sure way of bringing back unpleasant memories of lower school.[4]

Following some brief and, as I was later to observe, characteristically tense small talk with Rosy, Maurice and I walked down the Strand and across to Choy's Restaurant in Soho. Maurice's mood was surprisingly jovial. Slowly and precisely he detailed how, in spite of much elaborate crystallographic analysis, little real progress had been made by Rosy since the day she arrived at King's. Though her X-ray photographs were somewhat sharper than his, she was unable to say anything more positive than he had already. True, she had done some more detailed measurements of the water content of her DNA samples, but even here Maurice had doubts about whether she was really measuring what she claimed.

To my surprise, Maurice seemed buoyed up by my presence. The aloofness that existed when we first met in Naples had vanished. The fact that I,

Part of the airmail letter Franklin wrote to Anne Sayre on March 1, 1952. In the section shown, she mentions how good the facilities are at King's, but how disappointing her colleagues.

a phage person, found what he was doing important was reassuring. It really was no help to receive encouragement from a fellow physicist. Even when he met those who thought his decision to go into biology made sense, he couldn't trust their judgment. After all, they didn't know any biology, and so it was best to take their remarks as politeness, even condescension, toward someone opposed to the competitive pace of postwar physics.

Party at King's. Wilkins third from left, Randall extreme right.

To be sure, he got active and very necessary help from some biochemists. If not, he could never have come into the game. Several of them had been absolutely vital in generously providing him with samples of highly purified DNA. It was bad enough learning crystallography without having to acquire the witchcraft-like techniques of the biochemist. On the other hand, the majority weren't like the high-powered types he had worked with on the bomb project. Sometimes they seemed even ignorant of the way DNA was important.

But even so they knew more than the majority of biologists. In England, if not everywhere, most botanists and zoologists were a muddled lot. Not even the possession of University Chairs gave many the assurance to do clean science; some actually wasted their efforts on useless polemics about the origin of life or how we know that a scientific fact is really correct. What was worse, it was possible to get a university degree in biology without learning any genetics. That was not to say that the geneticists themselves provided any intellectual help. You would have thought that with all their talk about genes they should worry about what they were. Yet almost none of them seemed to take seriously the evidence that genes were made of DNA. This fact was unnecessarily chemical.[5] All that most of them wanted out of life was to set their students onto uninterpretable details of chromosome be-

[5] As late as 1955, C. D. Darlington, the eminent geneticist, would write: "According to Watson and Crick, the DNA normally exists in the chromosome as a double and coiled nucleotide column, each side of which, if separated from its partner, can act as a template for the assembling of the other. On this basis DNA would appear as a self-sufficient genetic structure with the protein at least mechanically subordinate. But, of course, we do not need to take such an extreme view: equality and reciprocity are also conceivable."

The Brains Trust: *Margery Fry, Julian Huxley, Commander Campbell, Robert Graves, C. E. M. Joad, and W. D. H. McCullough discuss questions sent in by listeners.*

[6] A number of noted scientists regularly discussed social, political, and intellectual topics on the radio and in newsprint. Julian Huxley was a major presence on the BBC radio (and then television) program *The Brains Trust*, on which panelists debated ideas in response to listeners' questions. J. B. S. Haldane wrote regularly in the pages of the communist newspaper *The Daily Worker*, while J. D. Bernal wrote many articles and books for the general public on both political and scientific topics, including the origin of life.

havior or to give elegantly phrased, fuzzyminded speculations over the wireless on topics like the role of the geneticist in this transitional age of changing values.[6]

So the knowledge that the phage group took DNA seriously made Maurice hope that times would change and he would not have painfully to explain, each time he gave a seminar, why his lab was making so much fuss and bother about DNA. By the time our dinner was finished, he was clearly in a mood to push ahead. Yet all too suddenly Rosy popped back into the conversation, and the possibility of really mobilizing his lab's efforts slowly receded as we paid the bill and went out into the night.

Chapter 11

[1] At the time of Watson and Crick's visit, Hodgkin's search for the structure of insulin had been going on for 15 years, and it would be more than another 15 years before she succeeded (in 1969). Before then, she would win the Nobel Prize in Chemistry (1964) for solving the structure of other important molecules including penicillin and vitamin B12. Her mentor and sometime lover had been J. D. Bernal (Sage) whose well-known communist sympathies she too embraced at this time.

[2] Crick was indeed misled about the water content "…partly due to Watson's misunderstanding of a technical crystallographic term Rosalind had used…He mixed up 'asymmetric unit' and 'unit cell'." But also, as Crick pointed out, he should himself have realized that the water content Watson recollected was too low, and that a sodium ion would be highly hydrated.

The following morning I joined Francis at Paddington Station. From there we were to go up to Oxford to spend the weekend. Francis wanted to talk to Dorothy Hodgkin, the best of the English crystallographers, while I welcomed the opportunity to see Oxford for the first time. At the train gate Francis was in top form. The visit would give him the opportunity to tell Dorothy about his success with Bill Cochran in working out the helical diffraction theory. The theory was much too elegant not to be told in person—individuals like Dorothy who were clever enough to understand its power immediately were much too rare.[1]

Dorothy Hodgkin with J. D. Bernal, 1937.

As soon as we were in the train carriage, Francis began asking questions about Rosy's talk. My answers were frequently vague, and Francis was visibly annoyed by my habit of always trusting to memory and never writing anything on paper. If a subject interested me, I could usually recollect what I needed. This time, however, we were in trouble, because I did not know enough of the crystallographic jargon. Particularly unfortunate was my failure to be able to report exactly the water content of the DNA samples upon which Rosy had done her measurements. The possibility existed that I might be misleading Francis by an order-of-magnitude difference.[2]

The wrong person had been sent to hear Rosy. If Francis had gone along, no such ambiguity would have existed. It was the penalty for being over-sensitive to the situation. For, admittedly, the sight of Francis mulling over the consequences of Rosy's information when it was hardly out of her mouth would have upset Maurice. In one sense it would be grossly unfair for them to learn the facts at the same time. Certainly Maurice should have the first chance to come to grips with the problem. On the other hand, there seemed no indication that he thought the answer would come from playing with molecular models. Our conversation on the previous night had hardly alluded to that approach. Of course, the possibility existed that he was keeping something back. But that was very unlikely—Maurice just wasn't that type.

The only thing that Francis could do immediately was to seize the water value, which was easiest to think about. Soon something appeared to make sense, and he began scribbling on the vacant back sheet of a manuscript he had been reading. By then I could not understand what Francis was up to and reverted to *The Times* for amusement. Within a few minutes, however, Francis made me lose all interest in the outside world by telling me that only a small number of formal solutions were compatible both with the Cochran-Crick theory and with Rosy's experimental data. Quickly he began to draw more diagrams to show me how simple the problem was. Though the math-

The masthead of that day's copy of The Times *(November 22, 1951). The front page was entirely given over to advertisements, a well-known feature of the newspaper until May 1966.*

ematics eluded me, the crux of the matter was not difficult to follow. Decisions had to be made about the number of polynucleotide chains within the DNA molecule. Superficially, the X-ray data were compatible with two, three, or four strands. It was all a question of the angle and radii at which the DNA strands twisted about the central axis.

By the time the hour-and-a-half train journey was over, Francis saw no reason why we should not know the answer soon. Perhaps a week of solid fiddling with the molecular models would be necessary to make us absolutely sure we had the right answer. Then it would be obvious to the world that Pauling was not the only one capable of true insight into how biological molecules were constructed. Linus' capture of the α-helix was most embarrassing for the Cambridge group. About a year before that triumph, Bragg, Kendrew and Perutz had published a systematic paper on the conformation of the polypeptide chain, an attack that missed the point.[3] Bragg in fact was

---

Polypeptide chain configurations in crystalline proteins

BY SIR LAWRENCE BRAGG, F.R.S., J. C. KENDREW AND M. F. PERUTZ

Cavendish Laboratory, University of Cambridge

(*Received* 31 *March* 1950)

Astbury's studies of α-keratin, and X-ray studies of crystalline haemoglobin and myoglobin by Perutz and Kendrew, agree in indicating some form of folded polypeptide chain which has a repeat distance of about 5·1 Å, with three amino-acid residues per repeat. In this paper a systematic survey has been made of chain models which conform to established bond lengths and angles, and which are held in a folded form by N—H—O bonds. After excluding the models which depart widely from the observed repeat distance and number of residues per repeat, an attempt is made to reduce the number of possibilities still further by comparing vector diagrams of the models with Patterson projections based on the X-ray data. When this comparison is made for two-dimensional Patterson projections on a plane at right angles to the chain, the evidence favours chains of the general type proposed for α-keratin by Astbury. These chains have a dyad axis with six residues in a repeat distance of 10·2 Å, and are composed of approximately coplanar folds. As a further test, these chains are placed in the myoglobin structure, and a comparison is made between calculated and observed F values for a zone parallel to the chains; the agreement is remarkably close taking into account the omission from the calculations of the unknown effect of the side-chains. On the other hand, a study of the three-dimensional Patterson of haemoglobin shows how cautious one must be in accepting this agreement as significant. Successive portions of the rod of high vector density which has been supposed to represent the chains give widely different projections and show no evidence of a dyad axis.

The evidence is still too slender for definite conclusions to be drawn, but it indicates that a further intensive study of these proteins, and in particular of myoglobin which has promising features of simplicity, may lead to a determination of the chain structure.

21-2

---

[3] Bragg et al.'s misjudged paper on polypeptide configurations. The authors were led astray because they assumed that there had to be an integral number of peptide units per turn. Furthermore, they did not make the fact that the peptide bond is planar a central element of their models (*Proc Roy Soc A* 203: 321–357).

still bothered by the fiasco. It hurt his pride at a tender point. There had been previous encounters with Pauling, stretching over a twenty-five-year interval. All too often Linus had got there first.

Even Francis was somewhat humiliated by the event. He was already in the Cavendish when Bragg had become keen about how a polypeptide chain folded up. Moreover, he was privy to a discussion in which the fundamental blunder about the shape of the peptide bond was made. That had certainly been the occasion to interject his critical facility in assessing the meaning of experimental observations—but he had said nothing useful. It was not that Francis normally refrained from criticizing his friends. In other instances he had been annoyingly candid in pointing out where Perutz and Bragg had publicly overinterpreted their hemoglobin results. This open criticism was certainly behind Sir Lawrence's recent outburst against him. In Bragg's view, all that Crick did was to rock the boat.[4]

Now, however, was not the time to concentrate on past mistakes. Instead, the speed with which we talked about possible types of DNA structures gathered intensity as the morning went by. No matter in whose company we found ourselves, Francis would quickly survey the progress of the past few hours, bringing our listener up to date on how we had decided upon models in which the sugar-phosphate backbone was in the center of the molecule. Only in that way would it be possible to obtain a structure regular enough to give the crystalline diffraction patterns observed by Maurice and Rosy. True, we had yet to deal with the irregular sequence of the bases that faced the outside—but this difficulty might vanish in the wash when the correct internal arrangement was located.

There was also the problem of

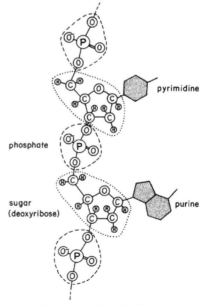

Sugar-phosphate backbone.

4 At a July 1951 seminar held at the Cavendish and entitled "What Mad Pursuit," Crick dismissed most of the approaches being used by his colleagues to solve the structure of hemoglobin. As reported in his later memoir of the same title, Crick says of his performance:

"Bragg was furious. Here was this newcomer telling experienced X-ray crystallographers, including Bragg himself who had founded the subject and been in the forefront of it for almost forty years, that what they were doing was most unlikely to lead to any useful result. The fact that I clearly understood the theory of the subject and indeed was apt to be unduly loquacious about it did not help."

what neutralized the negative charges of the phosphate groups of the DNA backbone. Francis, as well as I, knew almost nothing about how inorganic ions were arranged in three dimensions. We had to face the bleak situation that the world authority on the structural chemistry of ions was Linus Pauling himself. Thus if the crux of the problem was to deduce an unusually clever arrangement of inorganic ions and phosphate groups, we were clearly at a disadvantage. By midday it became imperative to locate a copy of Pauling's classic book, *The Nature of the Chemical Bond*.[5] Then we were hav-

[5] *The Nature of the Chemical Bond*, first published in 1939, integrated for the first time ideas of quantum mechanics into thinking about the chemical bond. It became an instant classic, widely consulted by professional scientists as well as being used as a text for high-level chemistry college courses.

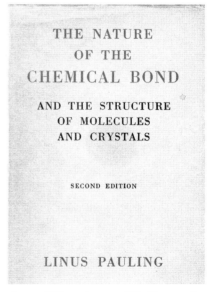

Cover of Watson's copy of **The Nature of the Chemical Bond.**

Manuscript page of **The Nature of the Chemical Bond.**

Blackwell's bookshop, 1950s.

[6] This was Watson's introduction to the Mitchison family. Upon his return to Cambridge he enthusiastically reported this meeting to his sister in a letter of November 28, 1951:

"I was in Oxford for a very enjoyable weekend. I stayed with a young zoologist who is a fellow of Magdalen College. My friend is a nephew of J. B. S. Haldane and almost as bright as his famed uncle. His mother is a very successful novelist while the father is a Labour MP. Apparently the family is still very rich. They have a large mansion in Scotland. There is a chance I may be invited for Christmas."

ing lunch near High Street. Wasting no time over coffee, we dashed into several bookstores until success came in Blackwell's. A rapid reading was made of the relevant sections. This produced the correct values for the exact sizes of the candidate inorganic ions, but nothing that could help push the problem over the top.

When we reached Dorothy's lab in the University Museum, the manic phase had almost passed. Francis ran through the helical theory itself, devoting only a few minutes to our progress with DNA. Most of the conversation centered instead on Dorothy's recent work with insulin. Since darkness was coming on, there seemed no point in wasting more of her time. We then moved on to Magdalen, where we were to have tea with Avrion Mitchison and Leslie Orgel, both then fellows of the college.[6] Over cakes Francis was ready

Nicholas Avrion (Av) Mitchison, 1957.

Magdalen College, Oxford.

[7] Watson told Delbrück of his dreams to be an Oxford don (December 9, 1951). "At Oxford, I have stayed at Magdalen College and have had the experience of eating in the senior common room where no one can speak at breakfast. Drinking port after dinner at High Table is an experience difficult to describe but immensely interesting to participate in."

to talk about trivial things, while I quietly thought how splendid it would be if I could someday live in the style of a Magdalen don.[7]

Dinner with claret, however, restored the conversation to our impending triumph with DNA. By then we had been joined by Francis' close friend, the logician George Kreisel, whose unwashed appearance and idiom did not fit into my picture of the English philosopher. Francis greeted his arrival with great gusto, and the sound of Francis' laughter and Kreisel's Austrian accent dominated the spiffy atmosphere of the restaurant along High Street at which Kreisel had directed us to meet him. For a while Kreisel held forth on a way to make a financial killing by shifting money between the politically divided parts of Europe.[8]

Avrion Mitchison then rejoined us, and the conversation for a short

Kreisel (left) and friend.

[8] Crick first met Kreisel while both were engaged in secret wartime work at the Admiralty. Like Crick, Kreisel was a fierce intellect and precise debater who had no patience for woolly thinking or unnecessary politeness ("Are people stupid because they are polite, or polite because they are stupid?"). A long time friend of Iris Murdoch, and as a student admired by Wittgenstein, Kreisel encouraged Crick to explore adventures in his private as well as intellectual life; the two remained close friends for the next 50 years.

time reverted to the casual banter of the intellectual middle class. This sort of small talk, however, was not Kreisel's meat, and so Avrion and I excused ourselves to walk along the medieval streets toward my lodgings. By then I was pleasantly drunk and spoke at length of what we could do when we had DNA.

Chapter 12

[1] Kendrew had married Elizabeth, the widow of a close friend who was killed in World War II, in 1948. Elizabeth was a physician who qualified in 1951. The Kendrews divorced in 1956.

I gave John and Elizabeth Kendrew the scoop about DNA when I joined them for breakfast on Monday morning.[1] Elizabeth appeared delighted that success was almost within our grasp, while John took the news more calmly. When it came out that Francis was again in an inspired mood and I had nothing more solid to report than enthusiasm, he became lost to the sections of *The Times* which spoke about the first days of the new Tory government. Soon afterward, John went off to his rooms in Peterhouse, leaving Elizabeth and me to digest the implications of my unanticipated luck. I did not remain long, since the sooner I could get back to the lab, the quicker we could find out which of the several possible answers would be favored by a hard look at the molecular models themselves.

This is the headline in The Times *of October 30, reporting the early days of the new Tory government. In the General Election of October 23, 1951, the Labour government of Clement Attlee had been ousted and Sir Winston Churchill became Prime Minister for the second time.*

Both Francis and I, however, knew that the models in the Cavendish would not be completely satisfactory. They had been constructed by John some eighteen months before, for the work on the three-dimensional shape of the polypeptide chain. There existed no accurate representations of the groups of atoms unique to DNA. Neither phosphorus atoms nor the purine and pyrimidine bases were on hand. Rapid improvisation would be necessary since there was no time for Max to give a rush order for their construction. Making brand-new models might take all of a week, whereas an answer was possible within a day or so. Thus as soon as I got to the lab I began adding bits of copper wire to some of our carbon-atom models, thereby changing them into the larger-sized phosphorus atoms.[2]

Much more difficulty came from the necessity to fabricate representations of the inorganic ions. Unlike the other constituents, they obeyed no simple-minded rules telling us the angles at which they would form their respective

Sven Furberg, 1950.

[2] **Watson refers to Sven Furberg's determination of the structure of cytidine in the caption of this figure (*right*). Published in *Nature* in 1949, it is recognized as a tour-de-force. Furberg was a Norwegian physical chemist who spent two years with Bernal at Birkbeck College.**

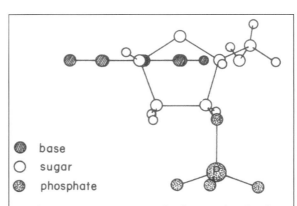

base

sugar

phosphate

A schematic view of a nucleotide, showing that the plane of the base is almost perpendicular to the plane in which most of the sugar atoms lie. This important fact was established in 1949 by S. Furberg, then working in London at J. D. Bernal's Birkbeck College lab. Later he built some very tentative models for DNA. But not knowing the details of the King's College experiments, he built only single-stranded structures, and so his structural ideas were never seriously considered in the Cavendish.

22　　　　　　　NATURE　　　July 2, 1949　Vol. 164

Wolfenstein. The rates obtained by Lewis and by Williams at the higher altitude for a chamber separation of 10 metres and for showers of the same density are also much greater than the predicted rate, by factors of 280 and 12 respectively. Part of the discrepancy may well be attributed to Wolfenstein's inclusion of the zenith angle effect, whereas recent experiments[2] indicate that extensive showers penetrating to the lower altitudes are incident nearly vertically.

Further details of the present work are to appear in the *Australian Journal of Scientific Research*, Series A, **2** (1949).

J. R. PRESCOTT
C. B. O. MOHR

Physics Department,
University of Melbourne.
March 17.

[1] Lewis, L. G., *Phys. Rev.*, **67**, 228 (1945).
[2] Williams, R., *Phys. Rev.*, **74**, 1689 (1948).
[3] Euler, H., *Z. Phys.*, **116**, 73 (1940).
[4] Wolfenstein, L., *Phys. Rev.*, **67**, 238 (1945).
[5] Carmichael, H., *Phys. Rev.*, **74**, 1667 (1948).
[6] Bridge, H., Hazen, W., Rossi, B., and Williams, R., *Phys. Rev.*, **74**, 1083 (1948).
[7] Bernadini, G., Cortini, C., and Manfredini, A., *Phys. Rev.*, **74**, 845 (1948).
[8] Montgomery, C. G., and Montgomery, D. D., *Phys. Rev.*, **72**, 131 (1947).
[9] Alichanian, A., and Asatiani, T., *J. Phys. U.S.S.R.*, **9**, 175 (1945).

Crystal Structure of Cytidine

A STUDY of the crystal structure of cytidine is being carried out by X-ray analysis. The crystal specimens were kindly supplied by Dr. D. O. Jordan, University of Nottingham, and were found to be orthorhombic with {110} dominating. An optical investigation shows that the sign is positive, with $\alpha \| c$, $\beta \| b$ and $\gamma \| a$. Cell dimensions are : $a = 13 \cdot 93$ A., $b = 14 \cdot 75$ A., $c = 5 \cdot 10$ A. ; density, $1 \cdot 53$; four molecules per unit cell ; space-group, $P\,2_1 2_1 2_1$.

Cytidine

Weissenberg photographs were taken, approximate atomic co-ordinates postulated by trial and error, and the Fourier map of the 001-projection shown in Fig. 1 eventually obtained. This map is now being refined.

Fig. 2 gives the interpretation of the peaks. The chemical formula is fully confirmed, thus showing cytidine to be cytosine-3-*d*-ribofuranoside. The glycosidic linkage is of the β-type, in accordance with the findings of Davoll, Lythgoe and Todd[1]. The bond-angles to the atom C_1' of the five-membered ring are not far from the tetrahedral angle, and the planes of the two ring systems are nearly perpendicular to each other. Details of the structure cannot be given at this stage ; but the pyrimidine-ring appears

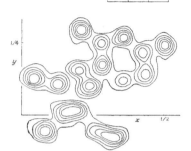

Fig. 1. Fourier projection of cytidine in direction of *c*-axis

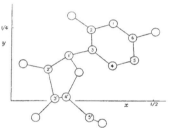

Fig. 2. Molecular projection corresponding to Fig. 1

to be flat, and there is some indication that the ribose-ring may not be planar.

Neighbouring molecules in the crystal are held together by hydrogen bonds.

It is hoped to publish later a more detailed account of the structure.

X-ray work on cytidylic acid is also in progress. The cell-dimensions are $a = 8 \cdot 74$ A., $b = 21 \cdot 4$ A., $c = 6 \cdot 82$ A., and the space-group $P\,2_1 2_1 2$.

S. FURBERG

Birkbeck College Research Laboratory,
21 Torrington Square,
London, W.C.1.

[1] Davoll, Lythgoe and Todd, *J. Chem. Soc.*, 833 (1946).

Runge Bands of O_2 in Flame Spectra

IN a recent note, Hornbeck[1] has reported observing the Runge emission bands of O_2, $^3\Sigma_u^- \rightarrow \,^3\Sigma_g^-$, in the spectra of explosion flames of carbon monoxide and oxygen, and has shown that this banded structure is favoured relative to the continuous background by excess of oxygen in the mixture. In a diffusion flame of carbon monoxide burning in oxygen at atmospheric pressure, we have confirmed the presence of the Runge bands, the (0,13), (0,14) and (0,15) bands, with heads at 3233, 3370 and 3516 A., being conspicuous. These bands, however, are emitted by a different part of the flame from the main carbon monoxide flame spectrum, and it is clear that the

Reprinted from *Acta Crystallographica*, Vol. 3, Part 5, September 1950

PRINTED IN GREAT BRITAIN

Acta Cryst. (1950). **3**, 325

The Crystal Structure of Cytidine

By S. Furberg*

Birkbeck College Research Laboratory, 21 Torrington Square, London W.C. 1, England

(Received 28 January 1950)

Crystals of cytidine, $C_9H_{13}O_5N_3$, are orthorhombic with $a = 13.93$, $b = 14.75$, $c = 5.10$ A. The space group is $P2_12_12_1$, and the unit cell contains four molecules. Atomic co-ordinates were postulated by extensive use of structure-factor graphs, supported by chemical and physical data, and then refined by two-dimensional Fourier syntheses. Direct confirmation is given that cytidine may be described as cytosine-3-β-D-ribofuranoside. The pyrimidine ring is found to be planar, whereas the D-ribose ring is non-planar with one of its atoms lying about 0.5 A. from the plane containing the four remaining atoms. The central C–N bond lies in the plane of the pyrimidine ring and makes tetrahedral angles with the adjacent ring bonds in the D-ribose. Evidence is found for a weak intra-molecular hydrogen bond between the 5'-hydroxyl group in the D-ribose and a (CH) group in the pyrimidine.

Rosalind Franklin's copy of Furberg's definitive paper on the structure of cytidine. Furberg's barely legible note beneath his signature reads "Hope you have been able to interpret your beautiful fibre diagram of Na-thymonucleate." Unfortunately we do not know when Franklin received this reprint.

[3] **The novel was *A Perch in Paradise*, by Margaret Bullard (Hamish Hamilton, 1952). Bertrand Russell enjoyed it, too. In a letter to Bullard on April 10, 1952, he wrote:**

"If Cambridge is as you represent it, it must have become more amusing since I was an undergraduate, which was in the early '90s. In those days we were all strictly celibate, which cannot be said of your characters. I am finding your novel amusing and pleasant reading and am hoping that it gives a true picture of Cambridge life."

chemical bonds. Most likely we had to know the correct DNA structure before the right models could be made. I maintained the hope, however, that Francis might already be on to the vital trick and would immediately blurt it out when he got to the lab. Over eighteen hours had passed since our last conversation, and there was little chance that the Sunday papers would have distracted him upon his return to the Green Door.

His tenish entrance, however, did not bring the answer. After Sunday supper he had again run through the dilemma but saw no quick answer. The problem was then put aside for a rapid scanning of a novel on the sexual misjudgments of Cambridge dons. The book had its brief good moments, and even in its most ill-conceived pages there was the

A Perch in Paradise.

Herbert (Freddie) Gutfreund flanked by Crick and Watson, outside Clare College, 1952.

Hugh Huxley, Kendrew's Ph.D. student and Ann Cullis, Max Perutz' assistant, 1950s.

question of whether any of their friends' lives had been seriously drawn on in the construction of the plot.[3]

Over morning coffee Francis nonetheless exuded confidence that enough experimental data might already be on hand to determine the outcome. We might be able to start the game with several completely different sets of facts and yet always hit the same final answers. Perhaps the whole problem would fall out just by our concentrating on the prettiest way for a polynucleotide chain to fold up. So while Francis continued thinking about the meaning of the X-ray diagram, I began to assemble the various atomic models into several chains, each several nucleotides in length. Though in nature DNA chains are very long, there was no reason to put together anything massive. As long as we could be sure it was a helix, the assignment of the positions for only a couple of nucleotides automatically generated the arrangement of all the other components.

The routine assembly task was over by one, when Francis and I walked over to the Eagle for our habitual lunch with the chemist Herbert Gutfreund. These days John usually went to Peterhouse, while Max always cycled home. Occasionally John's student Hugh Huxley would join us, but of late he was finding it difficult to enjoy Francis' inquisitive lunchtime attacks. For just prior to my arrival in Cambridge, Hugh's decision to take up the problem of how

[4] The Rock of Gibraltar is the promontory, 1400 feet high, bordering Spain and guarding the entrance to the Mediterranean. It has been in British hands since the Treaty of Utrecht in 1713. Despite numerous sieges, the Rock has never been captured, hence the expression used by Watson.

[5] Eprime Eshag was an Iranian economist, an ardent follower of John Maynard Keynes, who came to Cambridge to work for a Ph.D. thesis on the history of monetary theory. He worked for the United Nations before joining Wadham College, Oxford. According to his obituary he was also "an unrepentant man of many girlfriends" who married late, in 1992, dying 6 years later at the age of 80.

muscles contract had focused Francis' attention on the unforeseen opportunity that, for twenty years or so, muscle physiologists had been accumulating data without tying them into a self-consistent picture. Francis found it a perfect situation for action. There was no need for him to ferret out the relevant experiments since Hugh had already waded through the undigested mass. Lunch after lunch, the facts were put together to form theories which held for a day or so, until Hugh could convince Francis that a result he would like ascribed to experimental error was as solid as the Rock of Gibraltar.[4] Now the construction of Hugh's X-ray camera was completed, and soon he hoped to get experimental evidence to settle the debatable points. The fun would be all lost if somehow Francis could correctly predict what he was going to find.

But there was no need that day for Hugh to fear a new intellectual invasion. When we walked into the Eagle, Francis did not exchange his usual raucous greetings with the Persian economist Ephraim Eshag, but gave the undistilled impression that something serious was up.[5] The actual model building would start right after lunch, and more concrete plans must be formulated to make the process efficient. So over our gooseberry pie we looked at the pros and cons of one, two, three, and four chains, quickly dismissing one-chain helices as incompatible with the evidence in our hands. As to the forces that held the chains together, the best guess seemed to be salt bridges in which divalent cations like Mg^{++} held together two or more phosphate groups. Admittedly there was no evidence that Rosy's samples contained any divalent ions, and so

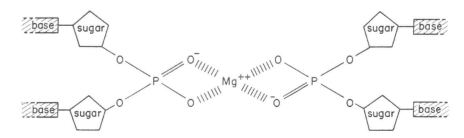

How Mg^{++} might be used to bind negatively charged phosphate groups in the center of a compound helix.

First page of Crick's memo on their triple helix model.

[6] This is the first page of a memorandum Crick wrote soon after Franklin's colloquium, setting out the principles which would guide Watson and him in devising a structure for DNA. In contrast to Franklin's view that experimental data were paramount, they would try "...to incorporate the *minimum* number of experimental facts" while acknowledging "...that certain results have suggested ideas to us." Crick emphasized that care had to be taken not to reject a model just "...because of some difficulty which will sort itself out at a later stage."

we might be sticking our necks out. On the other hand, there was absolutely no evidence against our hunch.[6] If only the King's groups had thought about models, they would have asked which salt was present and we would not be placed in this tiresome position. But, with luck, the addition of magnesium or possibly calcium ions to the sugar-phosphate backbone would quickly generate an elegant structure, the correctness of which would not be debatable.

Our first minutes with the models, though, were not joyous. Even though only about fifteen atoms were involved, they kept falling out of the awkward pincers set up to hold them the correct distance from one another. Even worse, the uncomfortable impression arose that there were no obvious restrictions on the bond angles between several of the most important atoms. This was not at all nice. Pauling had cracked the α-helix by ruthlessly following up his knowledge that the peptide bond was flat. To our annoyance, there seemed every reason to believe that the phosphodiester bonds which bound together the successive nucleotides in DNA might exist in a variety of shapes. At least with our level of chemical intuition, there was unlikely to be any single conformation much prettier than the rest.

After tea, however, a shape began to emerge which brought back our spirits. Three chains twisted about each other in a way that gave rise to a crystallographic repeat every 28 Å along the helical axis. This was a feature demanded by Maurice's and Rosy's pictures, so Francis was visibly reassured as he stepped back from the lab bench and surveyed the afternoon's effort. Admittedly a few of the atomic contacts were still too close for comfort, but, after all, the fiddling had just begun. With a few hours' more work, a presentable model should be on display.

Ebullient spirits prevailed during the evening meal at the Green Door. Though Odile could not follow what we were saying, she was obviously cheered by the fact that Francis was about to bring off his second triumph within the month. If this course of events went on, they would soon be rich and could own a car. At no moment did Francis see any point in trying to simplify the matter for Odile's benefit. Ever since she had told him that gravity went only three miles into the sky, this aspect of their relationship was set. Not only did she not know any science, but any attempt to put some in her head would be a losing fight against the years of her convent upbringing. The most to hope for was an appreciation of the linear way in which money was measured.

Our conversation instead centered upon a young art student then about to marry Odile's friend Harmut Weil. This capture was mildly displeasing to

Members of the M.R.C. Biophysics Unit at the annual Cricket match (1950s). From left to right: Maurice Wilkins, William (Willy) Seeds, Bruce Fraser, Mary Fraser, Ray Gosling (standing), Geoffrey Brown.

Francis. It was about to remove the prettiest girl from their party circle. Moreover, there was more than one thing cloudy about Harmut. He had come out of a German university tradition that believed in dueling.[7] There was also his undeniable skill in persuading numerous Cambridge women to pose for his camera.

All thought of women, however, was banished by the time Francis breezed into the lab just before morning coffee. Soon, when several atoms had been pushed in or out, the three-chain model began to look quite reasonable. The next obvious step would be to check it with Rosy's quantitative measurements. The model would certainly fit with the general locations of the X-ray reflections, for its essential helical parameters had been chosen to fit the seminar facts I had conveyed to Francis. If it were right, however, the model would also accurately predict the relative intensities of the various X-ray reflections.

A quick phone call was made to Maurice. Francis explained how the helical diffraction theory allowed a rapid survey of possible DNA models, and that he and I had just come up with a creature which might be the answer

[7] Alexander Todd (pictured on page 46) went to Frankfurt in 1929 to do research for his Ph.D. He recounts in his autobiography how he attended a duel, at 5 a.m., the object of which was for each opponent to inflict a wound on the face of the other. Afterwards, the combatants and observers "despite the early hour consumed vast quantities of beer" at a nearby inn.

we were all awaiting. The best thing would be for Maurice immediately to come and look it over. But Maurice gave no definite date, saying he thought he might make it sometime within the week. Soon after the phone was put down, John came in to see how Maurice had taken the news of the breakthrough. Francis found it hard to sum up his reply. It was almost as if Maurice were indifferent to what we were doing.

In the midst of further fiddling that afternoon, a call came through from King's. Maurice would come up on the 10:10 train from London the following morning. Moreover, he would not be alone. His collaborator Willy Seeds would also come. Even more to the point was that Rosy, together with her student R. G. Gosling, would be on the same train. Apparently they were still interested in the answer.

Chapter 13

Maurice decided to take a cab from the station to the lab. Ordinarily he would have come by bus, but now there were four of them to share the cost. Moreover, there would be no satisfaction in waiting at the bus stop with Rosy. It would make the present uncomfortable situation worse than it need be.

His well-intentioned remarks never came off, and even now, when the possibility of humiliation hung over them, Rosy was as indifferent as ever to his presence and directed all her attention to Gosling. There was only the slightest effort made at a united appearance when Maurice poked his head into our lab to say they had come. Especially in sticky situations like this, Maurice thought that a few minutes without science was the way to proceed. Rosy, however, had not come here to throw out foolish words, but quickly wanted to know where things stood.

Ray Gosling, early 1950s.

Neither Max nor John did anything to take the stage away from Francis. This was his day, and after they came in to greet Maurice they both pleaded pressure of their work to retire behind the closed doors of their joint office. Before the delegation's arrival, Francis and I had agreed to reveal our progress in two stages. Francis would first sum up the advantages of the helical theory. Then together we could explain how we had arrived at the proposed model for DNA. Afterwards we could go to the Eagle for lunch, leaving the afternoon free to discuss how we could all proceed with the final phases of the problem.

The first part of the show ran on schedule. Francis saw no reason to understate the power of the helical theory and within several minutes revealed the way Bessel functions gave neat answers. None of the visitors, however,

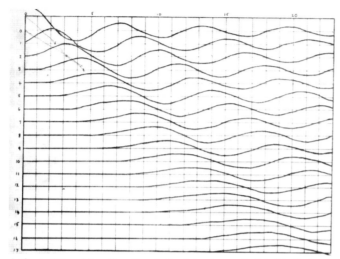

A chart of Bessel functions for a helical structure calculated by Stokes who called them "Waves at Bessel-on-Sea."

[1] Forty years later, Stokes recalled how he had come up with his version of helical diffraction theory: "So Maurice Wilkins, knowing my liking for mathematics and especially for Fourier analysis, said: 'Could you work out just what sort of an X-ray pattern would be given by a helical structure?' I said I thought I could. I mulled over this on the train home, and realized that the Fourier analysis that was needed for the problem would be full of Bessel functions. I was fortunate in that I had met Bessel functions before, in different contexts, so I knew what sort of creatures they were, and I was not at all scared of them. I came back next day with a diagram of some Bessel functions…which showed a remarkable likeness of the X-ray diffraction patterns that we had been getting. [These] came to be known as Waves at Bessel-on-Sea."

Stokes' calling his plot of Bessel functions "Bessel-on-Sea" was a pun on Bexhill-on-Sea, a seaside resort on the south coast of England popular for day-trips by train from London.

gave any indication of sharing Francis' delight. Instead of wishing to do something with the pretty equations, Maurice wanted to concentrate on the fact that the theory did not go beyond some mathematics his colleague Stokes had worked out without all this fanfare. Stokes had solved the problem in the train while going home one evening and had produced the theory on a small sheet of paper the next morning.[1]

Rosy did not give a hoot about the priority of the creation of the helical theory and, as Francis prattled on, she displayed increasing irritation. The sermon was unnecessary, since to her mind there was not a shred of evidence that DNA was helical. Whether this was the case would come out of further X-ray work. Inspection of the model itself only increased her disdain. Nothing in Francis' argument justified all this fuss. She became positively aggressive when we got on the topic of Mg^{++} ions that held together the phosphate groups of our three-chain model. This feature had no appeal at all to Rosy, who curtly pointed out that the Mg^{++} ions would be surrounded by tight shells of water molecules and so were unlikely to be the kingpins of a tight structure.[2]

Most annoyingly, her objections were not mere perversity: at this stage the embarrassing fact came out that my recollection of the water content of Rosy's

---

[2]*Raymond Gosling's account of the visit (2012)*

Maurice was telephoned by Francis to say that he and Jim Watson had built a helical model of DNA and would we all like to come up to Cambridge to look at it. Somewhat surprised, we took the train from Liverpool Street. It was a rather quiet journey due to the tension between Maurice and Rosalind and partly apprehension that the pair in the Cavendish might have scooped all our efforts. However, once in their Lab and in front of their model, our relief must have been palpable. Rosalind let rip in her best pedagogic style "you're wrong for the following reasons…" which she proceeded to enumerate as she demolished their proposal. Their mistake was to put the phosphate groups on the inside of their proposed structure. This gave otherwise wobbly unstable chains a strong central core. However, our diffraction patterns clearly showed that the phosphate groups, as the heaviest X-ray scatterers in the structure, must be on the outside of any molecule on a radius of about 10 Å around a central core containing the rest of the nucleotides. As Rosalind pointed out, this was confirmed by our experiments showing the ease with which water could pass in and out of the structure, whatever that might be.

It also confirmed Rosalind's view that one could build atomic models "until the cows came home" but it would be impossible to say which were nearer to the truth. If Maurice would back off and let us (she and I) get on with the measurement of the diffracted intensities and the admittedly slow and painstaking calculations, then ultimately "the data would speak for itself."

Rush hour at Liverpool Street station, London, October 12, 1951.

DNA samples could not be right. The awkward truth became apparent that the correct DNA model must contain at least ten times more water than was found in our model. This did not mean that we were necessarily wrong—with luck the extra water might be fudged into vacant regions on the periphery of our helix. On the other hand, there was no escaping the conclusion that our argument was soft. As soon as the possibility arose that much more water was involved, the number of potential DNA models alarmingly increased.

Though Francis could not help dominating the lunchtime conversation, his mood was no longer that of a confident master lecturing hapless colonial children who until then had never experienced a first-rate intellect. The group holding the ball was clear to all. The best way to salvage something from the day was to come to an agreement about the next round of experiments. In particular, only a few weeks' work should be necessary to see whether the DNA structure was dependent upon the exact ions used to neutralize the negative phosphate groups. Then the beastly uncertainty as to whether Mg^{++} ions were important could vanish. With this accomplished, another round of model building could start and, given luck, it might occur by Christmas.

Our subsequent after-lunch walk into King's and along the backs to Trinity did not, however, reveal any converts. Rosy and Gosling were pugnaciously assertive: their future course of action would be unaffected by their fifty-mile excursion into adolescent blather. Maurice and Willy Seeds gave more indication of being reasonable, but there was no certainty that this was anything more than a reflection of a desire not to agree with Rosy.

The situation did not improve when we got back to the lab. Francis did not want to surrender immediately, so he went through some of the actual details of how we went about the model building. Nonetheless, he quickly lost heart when it became apparent that I was the only one joining the conversation. Moreover, by this time neither of us really wanted to look at our model. All its glamor had vanished, and the crudely improvised phosphorus atoms gave no hint that they would ever neatly fit into something of value. Then when Maurice mentioned that, if they moved with haste, the bus might enable them to get the 3:40 train to Liverpool Street Station, we quickly said good-bye.

William Seeds (second from right) at a Physics Department party in the early 1950s. Ray Gosling is partly hidden behind Seeds.

Chapter 14

[1] We know now that there was also correspondence between Crick and Wilkins on this matter. The letters tell us much, not only about this episode but also about the personalities of Wilkins and Crick. The letters are presented on the following pages.

Rosy's triumph all too soon filtered up the stairs to Bragg. There was nothing to do but appear unperturbed as the news of the upset confirmed the fact that Francis might move faster if occasionally he would close his mouth. The consequences spread in a predictable fashion. Clearly this was the moment for Maurice's boss to discuss with Bragg whether it made sense for Crick and the American to duplicate King's heavy investment in DNA.[1]

Sir Lawrence had had too much of Francis to be surprised that he had again stirred up an unnecessary tempest. There was no telling where he would let loose the next explosion. If he continued to behave this way, he could easily spend the next five years in the lab without collecting sufficient data to warrant an honest Ph.D. The chilling prospect of enduring Francis throughout the remaining years of his tenure as the Cavendish Professor was too much to ask of Bragg or anyone with a normal set of nerves. Besides, for too long he had lived under the shadow of his famous father, with most people falsely thinking that his father, not he, was responsible for the sharp insight behind Bragg's Law. Now when he should

Sir Lawrence and Lady Alice Bragg, 1951.

be enjoying the rewards accorded the most prestigious chair in science, he had to be responsible for the outrageous antics of an unsuccessful genius.

The decision was thus passed on to Max that Francis and I must give up DNA. Bragg felt no qualms that this might impede science, since inquiries to Max and John had revealed nothing original in our approach. After Pauling's success, no one could claim that faith in helices implied anything but an un-

An exchange of letters followed the debacle of the triple helix model. Here, in a rather formal letter, Wilkins informs Crick of the position of the King's College group: "I am afraid the average vote of opinion here, most reluctantly and with many regrets, is against your proposal to continue to work on n.a. [nucleic acid] in Cambridge." Wilkins also worries that he has been too open with Crick: "I personally feel that I have much to gain by discussing my own work with you and after your attitude on Saturday begin to have *very slight* [interpolated] uneasy feelings in this respect." Wilkins is giving a copy of the letter to Randall and suggests that Crick show the letter to Perutz.

BIOPHYSICS RESEARCH UNIT,
KING'S COLLEGE,
STRAND,
LONDON, W.C.2.
TELEPHONE: TEMPLE BAR 5651

Dr. F. Crick,
Cavendish Laboratory,
Free School Lane,
Cambridge   11th December 1951

My dear Francis,

Firstly, I want to say I was very sorry to rush off on Saturday without seeing you again and thanking you for the pleasant time.

I am afraid the average vote of opinion here, most reluctantly and with many regrets, is against your proposal to continue the work on n.a. in Cambridge. An argument here is put forward to show that your ideas are derived directly from statements made in the colloquium and this seems to me as convincing as your own argument that your approach is quite out of the blue. It is also said that your type of solution would in any case be arrived at here as our programme is followed through. Fraser is, however, very keen on the whole approach along your lines and has been especially so since your suggestions of a month ago.

Apart from this, I think it most important that an understanding be reached such that all members of our laboratory can feel in future, as in the past, free to discuss their work and interchange ideas with you and your laboratory. We are two M.R.C. Units and two Physics Departments with many connections. I personally feel that I have much to gain by discussing my own work *very slight* with you and after your attitude on Saturday begin to have uneasy feelings in this respect. Whatever the precise rights or wrongs of the case I think it most important to preserve good inter-lab relations.

If you and Jim were working in a laboratory remote from ours our attitude would be that you should go right ahead. I think it best to abide by the view taken by the majority of the structure people here and your Unit as a whole. If your Unit thinks our suggestion selfish, or contrary to the interests as a whole of scientific advance, please let us know.

I suggest you show this letter to Max for his information, and having discussed the matter with Randall I am, at his request,

letting him have a copy.

Yours very sincerely,

Maurice.

MEDICAL RESEARCH COUNCIL

(possibly you might
like to show this
to John)

BIOPHYSICS RESEARCH UNIT,
KING'S COLLEGE,
STRAND,
LONDON, W.C.2.
TELEPHONE: TEMPLE BAR 5651

Dec 11. 51.

Dear Francis,

This is just to say
how bloody browned off I am
entirely & how rotten I feel about
it all & how entirely friendly I am
(though it may possibly appear differently).
We are really between forces which may
grind all of us into little pieces. So far
as your interests are concerned I do very
much suggest it is best to make some sacrifices
of for credit for ideas in this connection. You can
see how the wind is blowing when I say
that I had to restrain Randall from writing
to Bragg complaining about your behaviour. Needless
to say I did restrain him but so far as

your security with Bragg is concerned
it is probably much more important
to pipe down & build up the idea of
a quiet steady worker who never
creates 'situations' than to collect all
the credit for your excellent ideas at
the expense of good will.

And you see it does make
me also a bit confused about our discussion
if you get too interested in everything which
is important; where I say confused I
mean confused, I have given & am now
largely incapable of any logical thinking in
relation to polynucleotide chains or anything.

And poor Jim — May I shed a crocodile
& very confused tear? & send him my best wishes & regards &
friendly greetings to both of you & if you should have
any ill feeling about the part I have played I hope you
will tell me. Yours M   regards to John too!

Again from Wilkins to Crick, and again written on December 11, this letter is very much more informal and clearly not for the eyes of Randall or Perutz. Wilkin's anguish is evident:

"This is just to say how bloody browned off I am entirely & how rotten I feel about it all & how entirely friendly I am (though it may possibly appear differently). We are really between forces which may grind all of us into little pieces."

After giving Crick advice about how he should behave with Bragg, Wilkins continues:

"And you see it *does* make me a bit confused about our discussions if you get too interested in everything which is important…" before adding a special condolence for Watson: "And poor Jim—may I shed a crocodile and very confused tear?" But he ends sending "friendly greetings to both of you and if you *should* have any ill feeling about the part I have played I hope you will tell me."

On receipt of the previous two letters from Wilkins, Crick drafted a cavalier response to Wilkins: "...cheer up and take it from us that even if we kicked you in the pants it was between friends. We hope our burglary will at least produce a united front in your group!"

complicated brain. Letting the King's group have the first go at helical models was the right thing in any circumstance. Crick could then buckle down to his thesis task of investigating the ways that hemoglobin crystals shrink when they are placed in salt solutions of different density. A year to eighteen months of steady work might tell something more solid about the shape of the hemoglobin molecule. With a Ph.D. in his pocket Crick could then seek employment elsewhere.

No attempt was made to appeal the verdict. To the relief of Max and John, we refrained from publicly questioning Bragg's decision. An open outcry would reveal that our professor was completely in the dark about what the initials DNA stood for. There was no reason to believe that he gave it one hundredth the importance of the structure of metals, for which he took great delight in making soap-bubble models. Nothing then gave Sir Lawrence more pleasure than showing his ingenious motion-picture film of how bubbles bump each other.[2]

Our reasonableness did not arise, however, from a desire to keep peace with Bragg. Lying low made sense because we were up the creek with models based on sugar-phosphate cores. No matter how we looked at them, they smelled bad. On the day following the visit from King's, a hard look was

[2] Watson is unfairly dismissive of Bragg's interest in bubbles. In 1947, he and John Nye showed that bubbles on a liquid surface form closely packed rafts which exhibit behaviors similar to atoms in a metal. Bubble rafts continued to be useful in developing new insights but molecular dynamics is now used to do atomic simulations.

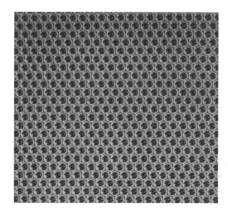

Perfect cystalline raft of bubbles. (Bragg and Nye, Proc Roy Soc A, *volume 190, plate 8.)*

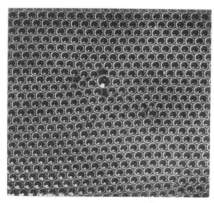

Effect of atoms of impurity. (Bragg and Nye, Proc Roy Soc A, *volume 190, plate 18.)*

given both to the ill-fated three-chain affair and to a number of possible variants. One couldn't be sure, but the impression was there that any model placing the sugar-phosphate backbone in the center of a helix forced atoms closer together than the laws of chemistry allowed. Positioning one atom the proper distance from its neighbor often caused a distant atom to become jammed impossibly close to its partners.

A fresh start would be necessary to get the problem rolling again. Sadly, however, we realized that the impetuous tangle with King's would dry up our source of new experimental results. Subsequent invitations to the research colloquia were not to be expected and even the most casual questioning of Maurice would provoke the suspicion that we were at it again. What was worse was the virtual certainty that cessation of model building on our part would not be accompanied by a burst of corresponding activity in their lab. So far, to our knowledge, King's had not built any three-dimensional models of the necessary atoms. Nonetheless, our offer to speed that task by giving them the Cambridge molds for making the models was only halfheartedly received. Maurice did say, though, that within a few weeks someone might be found to put something together, and it was arranged that the next time one of us went down to London the jigs could be dropped off at their lab.[3]

Thus the prospects that anyone on the British side of the Atlantic would crack DNA looked dim as the Christmas holidays drew near. Though Francis went back to proteins, obliging Bragg by working on his thesis was not to his liking. Instead, after a few days of relative silence, he began to spout about superhelical arrangements of the α-helix itself.[4] Only during the lunch hour could I be sure that he would talk DNA. Fortunately, John Kendrew sensed that the moratorium on working on DNA did not extend to thinking about it. At no time did he try to interest me in myoglobin. Instead, I used the dark and chilly days to learn more theoretical chemistry or to leaf through journals, hoping that possibly there existed a forgotten clue to DNA.

The book I poked open the most was Francis' copy of *The Nature of the Chemical Bond*. Increasingly often, when Francis needed it to look up a cru-

[3] A jig is used as a template for the reproduction of identical items. In this case the Cavendish workshop had made accurate representations of the four bases, and these jigs could be used to quickly make more of each base. Wilkins wrote later that Watson and Crick providing these jigs "...was a very good example of the way our science should have been moving" but that Franklin scorned their use.

[4] Crick's work on superhelical arrangements of the α-helix ("coiled coils") was to land him in yet more hot water in a priority dispute with Linus Pauling (Chapter 20).

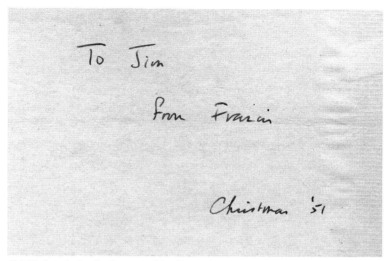

Crick's inscription in the copy of Pauling's **The Nature of the Chemical Bond,** *which he gave as a Christmas present to Watson.*

cial bond length, it would turn up on the quarter bench of lab space that John had given to me for experimental work. Somewhere in Pauling's masterpiece I hoped the real secret would lie. Thus Francis' gift to me of a second copy was a good omen. On the flyleaf was the inscription, "To Jim from Francis—Christmas '51." The remnants of Christianity were indeed useful.

Chapter 15

Naomi Mitchison, pictured near Carradale, 1955. Behind her is the fishing boat The Maid of Morven *which she owned jointly with local fisherman Denis McIntosh. Naomi Mitchison wrote science fiction and historical novels. She was a great friend of J. R. R. Tolkien and helped proofread* The Lord of the Rings.

I did not sit through the Christmas holidays in Cambridge. Avrion Mitchison had invited me to Carradale, the home of his parents, on the Mull of Kintyre. This was real luck, since over holidays Av's mother, Naomi, the distinguished writer, and his Labour MP father, Dick, were known to fill their large house with odd assortments of lively minds. Moreover, Naomi was a sister of England's most clever and eccentric biologist, J. B. S. Haldane. Neither the feeling that our DNA work had hit a roadblock nor the uncertainty of getting paid for the year was of much concern when I joined Av and his sister Val at Euston Station. No seats were left on the overnight Glasgow train, giving us a ten-hour journey seated on luggage listening to Val comment on the dull, boorish habits of the Americans who each year are deposited in increasing numbers at Oxford.

Gilbert (Dick) Mitchison campaigning as a Labour candidate in Kettering in the General Election of 1945. Mitchison won the seat, defeating the Conservative candidate John (Jack) Profumo. This image is taken from a British Council documentary film about that election.

J. B. S. Haldane exhorting the crowd at a United Front meeting in Trafalgar Square, 1937. The United Front was a coalition of workers from left-wing organizations formed to fight fascist Germany.

UNIVERSITY OF CAMBRIDGE   DEPARTMENT OF PHYSICS
CRYSTALLOGRAPHIC LABORATORY

TELEPHONE
CAMBRIDGE 55478

CAVENDISH LABORATORY
FREE SCHOOL LANE
CAMBRIDGE

Tuesday Afternoon

Dear Betty

To clarify our telephone conversation i) We will visit the home of Murdoch and Avrion Mitchison - They live at Carradale House, ~~Carra~~ Carradale. Kintyre, Scotland . General geography is as follows

it will take about 6 hours to go from glasgow, via train, Carradale. boat, and bus. The correct train leaves at 8:30 from Glasgow. Making connections

[1] *Letter from Watson to his sister describing how they would travel to Carradale. Watson was coming on the train from London with Av and his sister Val, meeting at Glasgow station Betty, who would be arriving by plane at Prestwick. The hand-drawn map of the "general geography" shows Carradale's location relative to Glasgow and Edinburgh.*

Elizabeth Watson, Cambridge, 1953.

At Glasgow we found my sister Elizabeth, who had flown to Prestwick from Copenhagen. Two weeks previously she had sent a letter relating that she was pursued by a Dane. Instantly I sensed impending disaster, for he was a successful actor. At once I inquired whether I could bring Elizabeth to Carradale. The affirmative reply I received with much relief, since it was inconceivable that my sister could think about settling in Denmark after two weeks of an eccentric country house.[1]

Dick Mitchison met the Campbeltown bus at the turnoff for Carradale to drive us the final twenty hilly miles to the tiny Scottish fishing village where he and Naomi had lived for the past twenty years. Dinner was still going on as we emerged from a stone passage, which connected the gunroom with several larders, into a dining room dominated by sharp authoritative chatter. Av's zoologist brother Murdoch had already come, and he enjoyed cornering people to talk about how cells divide. More often, the

Campbeltown bus, 1950.

theme was politics and the awkward cold war thought up by the American paranoids, who should be back in the law offices of middlewestern towns.

By the following morning I was aware that the best way not to feel impossibly cold was to remain in bed or, when that proved impossible, to go walking, unless the rain was coming down in buckets. In the afternoons Dick was always trying to get someone to shoot pigeons, but after one at-

Carradale House.

Avrion Mitchison as a child.

Naomi Mitchison, 1938.

Cover of Beyond This Limit.

[2] All the pictures above are by Wyndham Lewis, painter and writer, a founding member of the artistic movement known as the Vorticists, and editor of the literary magazine *BLAST*, which appeared briefly in 1914–1915. His novels include *Tarr* and *The Childermass*, while as an artist his portraits of T. S. Eliot and Edith Sitwell are particularly well known. In 1935, Lewis illustrated Naomi Mitchison's book, *Beyond This Limit*. Hemingway memorably described Lewis: "I have never seen a nastier man…Under the black hat, when I had first seen them, the eyes had been those of an unsuccessful rapist."

tempt, when I fired the gun after the pigeons were out of view, I took to lying on the drawing-room floor as close as possible to the fire. There was also the warming diversion of going to the library to play ping-pong beneath Wyndham Lewis' stern drawings of Naomi and her children.[2]

More than a week passed before I slowly caught on that a family of leftish leanings could be bothered by the way their guests were attired. Naomi and several of the women dressed for dinner, but I put this aberrant behavior down as a sign of approaching old age. The thought never occurred to me that my own appearance was noticed, since my hair was beginning to lose its American identity. Odile had been very shocked when Max introduced me to her on my first day in Cambridge and afterwards had told Francis that a bald American was coming to work in the lab. The best way to rectify the situation was to avoid a barber until I merged with the Cambridge scene. Though my sister was upset when she saw me, I knew that months, if not years, might be required to replace her superficial values with those of the English intellectual. Carradale thus was the perfect environment to go one step further and acquire a beard. Admittedly I did not like its reddish color, but shaving with cold water was agony. Yet after a week of Val's and Murdoch's acid comments, together with the expected unpleasantness of my sister, I emerged for dinner with a clean face. When Naomi made a complimentary remark about my looks, I knew that I had made the right decision.[3]

Murdoch Mitchison, 1963.

In the evenings there was no way to avoid intellectual games, which gave the greatest advantage to a large vocabulary. Every time my limpid contribution was read, I wanted to sink behind my chair rather than face the condescending stares of the Mitchison women. To my relief, the large number of house guests never permitted my turn to come often, and I made a point of sitting near the evening's box of chocolates, hoping no one would notice that I never passed it. Much more agreeable were the hours playing "Murder" in the dark twisting recesses of the upstairs floors. The most ruthless of the murder addicts was Av's sister Lois, then just back from teaching for a year in Karachi, and a firm proponent of the hypocrisy of Indian vegetable eaters.

Almost from the start of my stay I knew that I would depart from Naomi's and Dick's spectrum of the left with the greatest reluctance. The prospect of lunch with the alcoholic English cider more than compensated for the habit of leaving the outside doors open to the westerly winds. My departure, three days after the New Year, nonetheless had been fixed by Murdoch's arranging for me to speak at a London meeting of the Society for Experimental Biology. Two days before my scheduled departure there was a heavy fall of snow, giving to the barren moors the look of Antarctic mountains. It was a perfect occasion for a long afternoon walk along the closed Campbeltown Road, with Av talking about his thesis experiments on the transplantation of immunity while I thought about the possibility that the road might remain impassable through the day I was to leave. The climate was not with me, however, for a group from the house caught the Clyde steamer at Tarbert and the next morning we were in London.[4]

[3] Doris Lessing, Nobel Laureate in Literature, was a friend of Naomi Mitchison. In *Walking in the Shade*, the second volume of her autobiography, she describes staying at Carradale.

"An Incident: Naomi asked me to take a certain inarticulate young scientist for a walk. 'And for goodness' sake get him to say something—his tongue will atrophy.' His name was James Watson. For about three hours we walked about over the hills and through the heather, while I chatted away, my mother's daughter: one should know how to put people at their ease. At the end of it, exhausted, wanting only to escape, I at last heard human speech. 'The trouble is, you see, that there is only one other person in the world I can talk to.' I reported this to Naomi, and we agreed that it was as dandified a remark as we could remember, even from a very young man. Quite soon he and Francis Crick lay bare the structure of DNA."

Tarbert in Argyll and Bute.

[4] Writing to his parents after his return to Cambridge on January 8, 1952, Watson reported that: "All of [the Mitchisons] may be described as extreme individualist. As I am also somewhat of an extreme type, these people seemed quite 'normal' and I felt completely at ease."

Watson remained close to the Mitchisons. Nine days later (January 17, 1952) he wrote to his sister: "I went to a Trinity Feast with Murdoch. I'm losing all my principles since I was in dress clothes." In 1957, Watson was best man at Av and Lorna's wedding, and *The Double Helix* is dedicated to Naomi Mitchison.

Upon my return to Cambridge I had expected to hear from the States about my fellowship, but there was no official communication to greet me. Since Luria had written me in November not to worry, the absence of firm news by now seemed ominous. Apparently no decision had been made and

The wedding of Av Mitchison and Lorna Martin on the Isle of Skye. Naomi Mitchison, unidentified, Major-General Martin (Lorna's father), Watson's father, and Watson.

Naomi Mitchison and Watson on holiday in Antibes, Côte d'Azur, in 1958.

the worst was to be expected. The ax, however, could at most be only annoying. John and Max gave me assurance that a small English stipend could be dug up if I was completely cut off. Only in late January did my suspense end, with the arrival of a letter from Washington: I was sacked. The letter quoted the section of the fellowship award stating that the fellowship was valid only for work in the designated institution. My violation of this provision gave them no choice but to revoke the award.

The second paragraph gave the news that I had been awarded a completely new fellowship. I was not, however, to be let off merely with the long period of uncertainty. The second fellowship was not for the customary twelve-month period but explicitly terminated after eight months, in the middle of May. My real punishment in not following the Board's advice and going to Stockholm was a thousand dollars. By this time it was virtually impossible to obtain any support which could begin before the September start of a new school year. I naturally accepted the fellowship. Two thousand dollars was not to be thrown away.

Paul Weiss, new chairman of the Fellowship Board.

Less than a week later, a new letter came from Washington. It was signed by the same man, but not as head of the fellowship board. The hat he now displayed was that of the chairman of a committee of the National Research Council. A meeting was being arranged for which I was asked to give a lecture on the growth of viruses. The time of the meeting, to be held in Williamstown, was the middle of June, only a month after my fellowship would expire. I, of course, had not the slightest intention of leaving either in June or in September. The only problem was how to frame an answer. My first impulse was to write that I could not come because of an unforeseen financial disaster. But on second thought, I was against giving him the satisfaction of thinking he had affected my affairs. A letter went off saying that I found Cambridge intellectually very exciting and so did not plan to be in the States by June.[5]

[5] The fellowship saga continues (see Appendix 3). Luria, in replying to one of Watson's letters concerning his fellowship woes, alludes to Watson's burgeoning Anglophilia as well as his dislike of Paul Weiss, the Fellowship Board chairman: "As for Paul Weiss, I incline to agree with your definition, although being less British than you are, I would call him a 'damn son-of-a-bitch' rather than a 'bloody bastard'." (March 5, 1952)

Chapter 16

[1] Watson had turned to TMV "to mark time" after being forbidden to work on DNA, his frustration evident in a report to Delbrück (May 20, 1952):

"It is quite obvious that a great deal of work should go into elucidating this structure [DNA]. However the people at King's are involved in a fight among themselves and so at present no real effort is being made to solve the structure. We have attempted to interest them in a Pauling approach of model building and in fact this winter we did spend several weeks attempting to build plausible models. However, we have temporarily stopped for the political reason of not working on the problem of a close friend. If, however, the King's people persist in doing nothing, we shall again try out our luck."

By now I had decided to mark time by working on tobacco mosaic virus (TMV).[1] A vital component of TMV was nucleic acid, and so it was the perfect front to mask my continued interest in DNA. Admittedly the nucleic-acid component was not DNA but a second form of nucleic acid known as ribonucleic acid (RNA). The difference was an advantage, however, since Maurice could lay no claim to RNA. If we solved RNA we might also provide the vital clue to DNA. On the other hand, TMV was thought to have a molecular weight of 40 million and at first glance should be frightfully more difficult to understand than the much smaller myoglobin and hemoglobin molecules that John and Max had been working on for years without obtaining any biologically interesting answers.

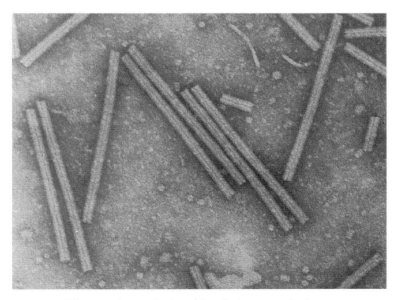

Electron micrograph of particles of tobacco mosaic virus.

Dorothy Hodgkin watches as Bernal and Fankuchen relax at the beach during the First Congress of the International Union of Crystallography, 1948.

[2] Bernal was one of the most outspoken and controversial of a group of left-wing scientists which was in the public eye from the 1930s through the 1950s. Bernal was an unrepentant Marxist, defending Lysenko and not being swayed by the horrors of Stalin's regime which were revealed by Khrushchev in the 1950s. In 1939, he wrote a controversial book, *The Social Function of Science,* arguing that scientific research had to be organized if all in society, and not just the elite, were to benefit. In the 1950s and 1960s, Bernal worked on the structure of water. He was notorious for his numerous affairs.

Moreover, TMV had previously been looked at with X rays by J. D. Bernal and I. Fankuchen. This in itself was scary, since the scope of Bernal's brain was legendary and I could never hope to have his grasp of crystallographic theory. I was even unable to understand large sections of their classic paper published just after the start of the war in the *Journal of General Physiology*. This was an odd place to publish; but Bernal had become absorbed in the war effort, and Fankuchen, by then returned to the States, decided to place their data in a journal looked at by people interested in viruses. After the war Fankuchen lost interest in viruses, and, though Bernal dabbled at protein crystallography, he was more concerned about furthering good relations with the Communist countries.[2]

Though the theoretical basis for many of their conclusions was shaky, the take-home lesson was obvious. TMV was constructed from a large number of identical subunits. How the subunits were arranged they did not know. Moreover, 1939 was too early to come to grips with the fact that the protein and RNA components were likely to be constructed along radically different lines. By now, however, protein subunits were easy to imagine in large numbers. Just the opposite was true of RNA. Division of the RNA component into a large number of subunits would produce polynucleotide chains too small to carry the genetic information that Francis and I believed must reside in the viral RNA. The most plausible hypothesis for the TMV structure was a central RNA core surrounded by a large number of identical small protein subunits.

In fact, there already existed biochemical evidence for protein building blocks. Experiments of the German Gerhard Schramm, first published in 1944, reported that TMV particles in mild alkali fell apart into free RNA and

Gerhard Schramm (front left), Rosalind Franklin, and Maurice Wilkins sit together for the group portrait at the 1956 Gordon Conference on Nucleic Acids and Proteins. Howard K. Schachman sits behind Schramm and Hamish N. Munro sits on the ground between Franklin and Wilkins.

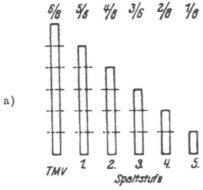

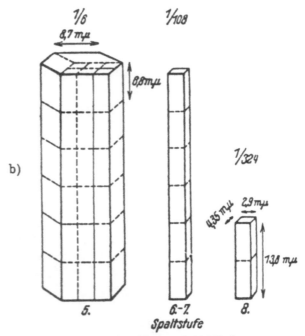

Schramm's figure illustrating how a TMV particle falls apart producing progressively shorter fragments and how these might be assembled.

a large number of similar, if not identical, protein molecules. Virtually no one outside Germany, however, thought that Schramm's story was right. This was because of the war. It was inconceivable to most people that the German beasts would have permitted the extensive experiments underlying his claims to be routinely carried out during the last years of a war they were so badly losing. It was all too easy to imagine that the work had direct Nazi support and that his experiments were incorrectly analyzed. Wasting time to disprove Schramm was not to most biochemists' liking. As I read Bernal's paper, however, I suddenly became enthusiastic about Schramm, for, if he had misinterpreted his data, by accident he had hit upon the right answer.[3]

Conceivably a few additional X-ray pictures would tell how the protein subunits were arranged. This was particularly true if they were helically stacked. Excitedly I pilfered Bernal's and Fankuchen's paper from the Philosophical Library and brought it up to the lab so that Francis could inspect the TMV X-ray picture. When he saw the blank regions that characterize heli-

[3] Schramm was part of a virus research group formed by researchers working in the Kaiser Wilhelm Institutes for Biochemistry and Biology in Dahlem. He was a member of the National Socialist German Workers Party (the Nazi party) but it seems that his research was never connected with the German war effort.

X-RAY AND CRYSTALLOGRAPHIC STUDIES OF PLANT VIRUS PREPARATIONS

I. INTRODUCTION AND PREPARATION OF SPECIMENS
II. MODES OF AGGREGATION OF THE VIRUS PARTICLES

By J. D. BERNAL AND I. FANKUCHEN*

(From the Department of Physics, Birkbeck College, University of London)

PLATES 1 TO 4

(Received for publication, March 14, 1941)

INTRODUCTION

Since their original isolation by Stanley in 1935, the protein preparations from plants suffering from virus diseases have been much studied, but chiefly chemically and biologically.   This paper is an account of a physical and crystallographic study of virus preparations which was carried out in conjunction with the work of Bawden and Pirie (1937 *a*, *b*; 1938 *a*, *b*, *c*).

The first of two papers by Bernal and Fankuchen describing their work on tobacco mosaic virus and other plant viruses, published in the Journal of General Physiology *25: 111–146.*

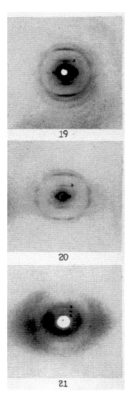

X-ray diffraction pictures of tobacco mosaic virus (19), cucumber mosaic virus (20), and potato virus X (21) from Bernal and Fankuchen's first paper.

cal patterns, he jumped into action, quickly spilling out several possible helical TMV structures. From this moment on, I knew I could no longer avoid actually understanding the helical theory. Waiting until Francis had free time to help me would save me from having to master the mathematics, but only at the penalty of my standing still if Francis was out of the room. Luckily, merely a superficial grasp was needed to see why the TMV X-ray picture suggested a helix with a turn every 23 Å along the helical axis. The rules were, in fact, so simple that Francis considered writing them up under the title, "Fourier Transforms for the Birdwatcher."

This time, however, Francis did not carry the ball and on subsequent days maintained that the evidence for a TMV helix was only so-so. My morale automatically went down, until I hit upon a foolproof reason why subunits should be helically arranged. In a moment of after-supper boredom I had read a Faraday Society Discussion on "The Structure of Metals." It contained an ingenious theory by the theoretician F. C. Frank on how crystals grow. Every time the calculations were properly done, the paradoxical answer emerged that the crystals could not grow at anywhere near the observed rates. Frank saw that the paradox vanished if crystals were not as regular as suspected, but contained dislocations resulting in the perpetual presence of cozy corners into which new molecules could fit.

Frederick Charles Frank, a theoretical physicist best known for his work on crystal growth and solid state physics. With R. V. Jones and others, he played a key role in Royal Air Force Intelligence during World War II.

Several days later, on the bus to Oxford, the notion came to me that each TMV particle should be thought of as a tiny crystal growing like other crystals through the possession of cozy corners. Most important, the simplest way to generate cozy corners was to have the subunits helically arranged. The idea was so simple that it had to be right. Every helical staircase I saw that weekend in Oxford made me more confident that other biological structures would also have helical symmetry. For over a week I pored over electron micrographs

Fig. 1 shows (a)
as a continuous deformation of a plane, (b) in a block model, the form of a
simple cubic crystal when a screw dislocation emerges normally at the

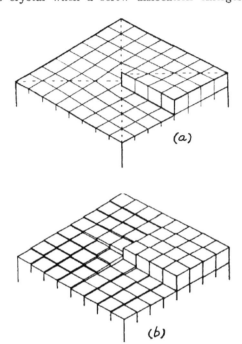

(a)

(b)

FIG. 1.—The end of a screw dislocation.

cube-face. It is clear that when dislocations of this type are present, the
crystal face always has exposed molecular terraces on which growth can
continue, and the need for fresh two-dimensional nucleation never arises.

*"Cozy corners," Watson's insight about the structure of TMV, came from F. C. Frank's 1949
paper "The influence of dislocations on crystal growth."* Discuss Faraday Soc 5: 48–54, p. 49.

of muscle and collagen fibers, looking for hints of helices. Francis, however,
remained lukewarm, and in the absence of any hard facts I knew it was futile to try to bring him around.

Hugh Huxley came to my rescue by offering to teach me how to set up
the X-ray camera for photographing TMV. The way to reveal a helix was to

tilt the oriented TMV sample at several angles to the X-ray beam. Fankuchen had not done this, since before the war no one took helices seriously. I thus went to Roy Markham to see if any spare TMV was on hand. Markham then worked in the Molteno Institute, which, unlike all other Cambridge labs, was well heated. This unusual state came from the asthma of David Keilin, then the "Quick Professor" and Director of Molteno. I always welcomed an excuse to exist momentarily at 70°F, even though I was never sure when Markham would start the conversation by saying how bad I looked, implying that if I had been brought up on English beer I would not be in my sorry state. This time he was unexpectedly sympathetic and without hesitation volunteered some virus. The idea of Francis and me dirtying our hands with experiments brought unconcealed amusement.

My first X-ray pictures revealed, not unexpectedly, much less detail than was found in the published pictures. Over a month was required before I could get even halfway presentable pictures. They were still a long way, though, from being good enough to spot a helix. The only real fun during February came from a costume party given by Geoffrey Roughton at his parents' home on Adams Road. Surprisingly, Francis did not wish to go, even though Geoffrey knew many pretty girls and was said to write poetry wearing one earring. Odile, however, did not want to miss it, so I went with her after hiring a Restoration soldier's garb. The moment we edged through the door into the crush of half-drunken dancers we knew the evening would be a smashing success, since seemingly half the attractive Cambridge *au pair* girls (foreign girls living with English families) were there.[4]

A week later there was a Tropical Night Ball that Odile was keen to attend, both since she had done the decorations and because it was sponsored by black people. Francis again demurred, this time wisely. The dance floor was half vacant, and even after several long drinks I did not enjoy dancing badly in open view. More to the point was that Linus Pauling was coming to London in May for a meeting organized by the Royal Society on the structure of proteins. One could never be sure where he would strike next. Particularly chilling was the prospect that he would ask to visit King's.[5]

[4] Geoffrey Roughton's enthusiasm for poetry involved him in a short-lived but popular Cambridge poetry magazine, called *Oasis*, in 1951–1952. He had written to Bragg that the goal of the magazine was to encourage more people, especially scientists, to read poetry. His father was Prof. F. J. W. Roughton, a physical chemist who had excited John Kendrew about proteins as an undergraduate at Trinity. His mother was Alice Roughton, a psychiatrist and medical campaigner who was known to have patients stay in her house to save them from institutionalization. Many parties were held in their house, including the one Watson attends at this point in the story. Watson recalls that Prof. Roughton ate a handkerchief as a party trick, and that on another occasion he arrived at their house to find a horse in the hall.

[5] The Royal Society's one-day discussion meeting was held on May 1, 1952. The importance of Pauling's presence at the meeting is highlighted by Astbury's remarks: "We are on the verge of great things in our understanding of the proteins, and one of our immediate tasks is to appraise these latest, most stimulating ideas put forward by Pauling and Corey."

Chapter 17

[1] Initially named after the golf course it displaced, Idlewild was renamed John F. Kennedy International Airport on December 24, 1963, one month after the assassination of President Kennedy.

Linus, however, was blocked from descending on London. His trip abruptly terminated at Idlewild through the removal of his passport.[1] The State Department did not want troublemakers like Pauling wandering about the globe saying nasty things about the politics of its onetime investment bankers who held back the hordes of godless Reds. Failure to contain Pauling might result in a London press conference with Linus expounding peaceful coexistence. Acheson's position was harassed enough without giving McCarthy the opportunity to announce that our government let radicals protected by U.S. passports set back the American way of life.

Francis and I were already in London when the scandal reached the Royal Society. The reaction was one of almost complete disbelief. It was far more reassuring to go on imagining that Linus had taken ill on the plane to

Idlewild Airport observation deck, late 1940s.

[2] Letter from Ruth B. Shipley, Chief of the Passport Division of the State Department, to Pauling, February 14, 1952. Addressing him as "My dear Dr. Pauling," she nevertheless denies his application for a passport "since the Department is of the opinion that your proposed travel would not be in the best interests of the United States." Shipley was chief of the division from 1928 until 1955, and wielded almost complete power over who got a passport and who didn't. Franklin Roosevelt described her as a "wonderful ogre," Secretary of State Dean Acheson said the Passport Division was her "Queendom of Passports," and in December 1951 *Time* magazine claimed she was "the most invulnerable, most unfirable, most feared and most admired career woman in Government."

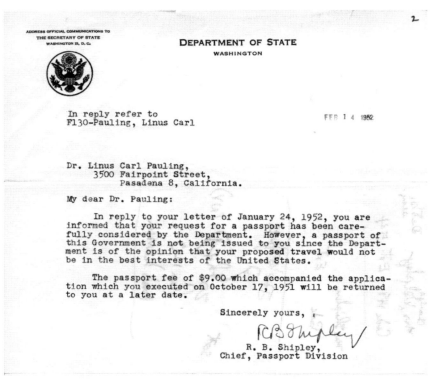

Letter from Shipley to Pauling.

New York. The failure to let one of the world's leading scientists attend a completely nonpolitical meeting would have been expected from the Russians. A first-rate Russian might easily abscond to the more affluent West. No danger existed, however, that Linus might want to flee. Only complete satisfaction with their Cal Tech existence came from him and his family.[2]

Several members of Cal Tech's governing board, however, would have been delighted with his voluntary departure. Every time they picked up a newspaper and saw Pauling's name among the sponsors of a World Peace Conference they seethed with rage, wishing there were a way to rid Southern California of his pernicious charm. But Linus knew better than to expect more than confused anger from the selfmade California millionaires

PAULING RAPS PASSPORT BAN

Caltech Man Denounces U. S. Refusal of Trip to London

"The damage done to the nation by the refusal to permit me to attend scientific meetings in England must be attributed to the McCarran Act, and is an argument for the repeal of this act."

Dr. Linus Pauling, California Institute of Technology chemistry department head, made this statement yesterday in discussing the State Department's refusal to issue him a passport, as disclosed in yesterday's Examiner.

He had accepted an invitation to speak before the Royal Society of London.

On April 28, despite an earlier appeal by letter to President Truman, officials of the State Department upheld an original refusal to issue him a passport on the ground "that my proposed travel would not be in the best interests of the United States."

Advised on April 21 by a State Department official that the decision was made because he was suspected of being a Communist, Dr. Pauling stated:

"I then submitted to the Department of State my statement, made under oath, that I am not a Communist, never have been a Communist, and never have been involved with the Communist Party, as well as other documents.

SCIENTIST—Dr. Linus Pauling stands beside a model showing protein structure that is the same as that in the horn he holds. He protested U. S. refusal to permit his attendance at London scientists' meeting where he was to explain his theories.
—Los Angeles Examiner photo

"The action of the State Department in refusing me a passport represents a different way of interfering with the progress of science and restricting the freedom of the individual citizen. In my opinion, it reflects a dangerous trend away from our fundamental democratic principles, upon which our nation is based."

Los Angeles Examiner *article.*

AN AMERICAN SCIENTIST

TO THE EDITOR OF THE TIMES

Sir,—On May 1 (possibly an unfortunate choice of day) the Royal Society held a significant symposium on the progress in our knowledge of proteins. As I had the honour to be President of the Royal Society when Professor Linus Pauling, of Pasadena, was awarded the Davy Medal (1947) and again when he was elected a foreign member (1948), it is perhaps appropriate that I should express the keen disappointment generally felt when it was learned that he had not been granted the necessary permit to make the journey to England in order to participate in the discussion. Pauling had an important contribution to make, and it is deplorable that we were deprived of the opportunity to talk it over with him.

It would be insincere to pretend that we have no inkling of the reason for the drastic action taken by the American authorities in this and several similar cases (e.g., that of Dr. E. B. Chain), but that does not lessen our surprise and consternation. It is an ironical circumstance that Pauling's theoretical views have been criticized in the U.S.S.R. as incorrect, western, and bourgeois; or, alternatively, as partly correct but anticipated by the Russian chemist Butlerow. To avoid any misunderstanding it must be added that I am not writing on behalf of the Royal Society.

Yours faithfully,
ROBERT ROBINSON.
The Dyson Perrins Laboratory, South Parks Road, Oxford, May 2.

Letter to The Times.

[3] The withholding of Pauling's passport received comment on both sides of the Atlantic. On the left, a piece printed in the *Los Angeles Examiner* from May 12, and, above, a letter from Sir Robert Robinson, former President of the Royal Society, to *The Times* on May 2. Probably because of such press attention, the State Department reversed its decision soon afterwards, allowing Pauling to travel later that summer (see Chapter 19).

whose knowledge of foreign policy was formed largely by the *Los Angeles Times*.[3]

The debacle was no surprise to several of us who had just been in Oxford for a Society of General Microbiology meeting on "The Nature of Viral Multiplication." One of the main speakers was to have been Luria. Two weeks prior to his scheduled flight to London, he was notified that he would

THE NATURE OF VIRUS MULTIPLICATION

SECOND SYMPOSIUM OF THE
SOCIETY FOR GENERAL MICROBIOLOGY
HELD AT OXFORD UNIVERSITY
APRIL 1952

CAMBRIDGE
Published for the Society for General Microbiology
AT THE UNIVERSITY PRESS
1953

CONTENTS

Although Luria's name and article appear in the published volume of the Symposium proceedings, he was refused a visa and did not attend the meeting.

SECRET   63a

P. F. 68,582/3.2.A./DLS.   7th August, 1953.

Dear Harlow,

Would you please refer to my letter
reference as above and dated 26th April, 1953.

We have now learned that Maurice
Hugh Frederick WILKINS is moving on 7th August,
1953 to 59 Great Cumberland Place.   I would
therefore be very grateful if you would arrange
for the H.O.W. at present operating on 184
Tottenham Court Road to be transferred to the
new address with effect from Saturday, 8th
August.

Yours sincerely,

D. L. Stewart.

G. A. Harlow, Esq.,
G.P.O.

DLS/slr.

One of many documents in Wilkins' MI5 files.

not get a passport. As usual, the State Department would not come clean
about what it considered dirt.[4]

Luria's absence thrust upon me the job of describing the recent experiments of the American phage workers. There was no need to put together a
speech. Several days before the meeting, Al Hershey had sent me a long letter
from Cold Spring Harbor summarizing the recently completed experiments by

[4] In a letter to his sister on April 3,
1952, Watson writes:

"I have just learned from mother that
Luria will not be coming. I do not
know the reason but I would suspect
passport difficulties. I am quite sorry
that he isn't coming since I had
hoped to discuss my future with him.
Now I shall have to do some involved
letter writing."

Left-leaning attitudes among scientists
weren't restricted to Pauling and
Luria, nor fear of them exclusive to the
U.S. authorities: Maurice Wilkins had
been investigated by MI5 and the FBI.
They suspected that one of nine scientists from New Zealand or Australia
had leaked A-bomb secrets. Wilkins
was one of the nine suspects, having
worked on the Manhattan project (see
Chapter 2). The investigation began in
1945, but was still going on in 1953, as
indicated by this request that the
H.O.W. (Home Office Warrant, which
allowed his mail to be searched) be
transferred to his new address. His
phone was also tapped. All this despite an informant suggesting around
the same time that while Wilkins was
"a very queer fish," he was probably a
socialist rather than a communist.

[5] François Jacob, then still a student with Lwoff, encountered Watson for the first time at this meeting at Oxford, recording the incident in his autobiography *The Statue Within.*

"At that time, to a French student who had not yet been inside an American university or seen its denizens, Jim Watson was an amazing character. Tall, gawky, scraggly, he had an inimitable style. Inimitable in his dress: shirttails flying, knees in the air, socks down around his ankles. Inimitable in his bewildered manner, his mannerisms: his eyes always bulging, his mouth always open, he uttered short, choppy sentences punctuated by 'Ah! Ah!' Inimitable also in his way of entering a room, cocking his head like a rooster looking for the finest hen, to locate the most important scientist present and charging over to his side. A surprising mixture of awkwardness and shrewdness. Of childishness in the things of life and of maturity in those of science."

Martha Chase and Al Hershey, 1953.

which he and Martha Chase established that a key feature of the infection of a bacterium by a phage was the injection of the viral DNA into the host bacterium. Most important, very little protein entered the bacterium. Their experiment was thus a powerful new proof that DNA is the primary genetic material.

Nonetheless, almost no one in the audience of over four hundred microbiologists seemed interested as I read long sections of Hershey's letter. Obvious exceptions were André Lwoff, Seymour Benzer, and Gunther Stent, all briefly over from Paris. They knew that Hershey's experiments were not trivial and that from then on everyone was going to place more emphasis on DNA. To most of the spectators, however, Hershey's name carried no weight. Moreover, when it came out that I was an American, my uncut hair provided no assurance that my scientific judgment was not equally bizarre.[5]

Dominating the meeting were the English plant virologists F. C. Bawden and N. W. Pirie. No one could match the smooth erudition of Bawden or the assured nihilism

Watson in shorts, Cold Spring Harbor, 1953.

Max Delbrück (left) with André Lwoff at Cold Spring Harbor, 1953.

F. C. Bawden.

N. W. Pirie (second from left).

of Pirie, who strongly disliked the notion that some phages have tails or that TMV is of fixed length. When I tried to corner Pirie about Schramm's experiments he said they should be dismissed, and so I retreated to the politically less controversial point of whether the 3000 Å length of many TMV particles was biologically important. The idea that a simple answer was preferable had no appeal to Pirie, who knew that viruses were too large to have well-defined structures.[6]

If it had not been for the presence of Lwoff, the meeting would have flopped totally. André was very keen about the role of divalent metals in phage multiplication and so was receptive to my belief that ions were decisively important for nucleic-acid structure. Especially intriguing was his hunch that specific ions might be the trick for the exact copying of macromolecules or the attraction between similar chromosomes. There was no way to test our dreams, however, unless Rosy did an about-face from her determination to rely completely on classical X-ray diffraction techniques.

At the Royal Society Meeting there was no hint that anyone at King's had mentioned ions since the confrontation with Francis and me in early December. Upon pressing Maurice, I learned that the jigs for the molecular models had not been touched after arriving at his lab. The time had not yet come to press Rosy and Gosling about building models. If anything, the

[6] Having met at Cambridge in the late 1920s, Frederick Bawden and Norman Pirie subsequently worked together for many years at Rothamsted Experimental Station at Harpenden. They collaborated first on potato virus X and then, starting in 1936, on TMV. They worked with Bernal and Fankuchen (see Chapter 16) on determining the chemical nature of TMV, and were the first to show the presence of RNA in the viral preparations. François Jacob's autobiography describes them at this meeting as "...old cronies who loved to play the buffoon, trading jokes and metaphysical aphorisms, all in a rapid, choppy English which left me in a cold sweat."

Pirie's later interests in a broad range of scientific and social issues are reflected in his picture above, which shows him leaving the Soviet Embassy with J. B. Priestly and others after talks on nuclear armament in September 1961.

squabbling between Maurice and Rosy was more bitter than before the visit to Cambridge. Now she was insisting that her data told her DNA was *not* a helix. Rather than build helical models at Maurice's command, she might twist the copper-wire models about his neck.

When Maurice asked whether we needed the molds back in Cambridge, we said yes, half implying that more carbon atoms were needed to make models showing how polypeptide chains turned corners. To my relief, Maurice was very open about what was not happening at King's. The fact that I was doing serious X-ray work with TMV gave him assurance that I should not soon again become preoccupied with the DNA pattern.

Watson in Paris on his way to the Riviera, spring 1952. He wrote to his sister (April 27): "I'm enclosing a photo of myself in Paris. It rather horrifies me since I did not realize how much hair I have. Needless to say I no longer have a crew cut."

Chapter 18

Maurice had no suspicion that almost immediately I would get the X-ray pattern needed to prove that TMV was helical. My unexpected success came from using a powerful rotating anode X-ray tube which had just been assembled in the Cavendish. This supertube permitted me to take pictures twenty times faster than with conventional equipment. Within a week I more than doubled the number of my TMV photographs.

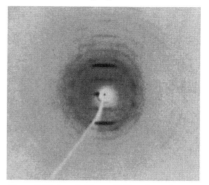

One of Watson's X-ray diffraction photographs of tobacco mosaic virus.

Custom then locked the doors of the Cavendish at 10:00 P.M. Though the porter had a flat next to the gate, no one disturbed him after the closing hour. Rutherford had believed in discouraging students from night work, since the summer evenings were more suitable for tennis. Even fifteen years after his death there was only one key available for late workers. This was now pre-empted by Hugh Huxley, who argued that muscle fibers were living and hence not subject to rules for physicists. When necessary, he lent me the key or walked down the stair to unlock the heavy doors that led out onto Free School Lane.

Hugh was not in the lab when late on a midsummer June night I went back to shut down the X-ray tube and to develop the photograph of a new TMV sample. It was tilted at about 25 degrees, so that if I were lucky I'd find the helical reflections. The moment I held the still-wet negative against the light box, I knew we had it. The telltale helical markings were unmistakable. Now there should be no problem in persuading Luria and Delbrück that my staying in Cambridge made sense. Despite the midnight hour, I had no desire to go back to my room on Tennis Court Road, and happily I walked along the backs for over an hour.

10   BIOCHIMICA ET BIOPHYSICA ACTA   VOL. **13** (1954)

THE STRUCTURE OF TOBACCO MOSAIC VIRUS

I. X-RAY EVIDENCE OF A HELICAL ARRANGEMENT OF SUB-UNITS AROUND THE LONGITUDINAL AXIS

by

J. D. WATSON*

The Medical Research Council Unit for the Study of the Molecular Structure of Biological Systems, Cavendish Laboratory, Cambridge (England)

INTRODUCTION

Tobacco Mosaic Virus (TMV) is a well known structure. It has a molecular weight of about $4 \cdot 10^7$ and is rod shaped with a diameter of 150 A and a length of 2800 A. Like other plant viruses, it contains both ribonucleic acid (6%) and protein (94%). In liquid solution, the virus may aggregate lengthwise and form regions of parallel orientation in which the particles are arranged in hexagonal close packing at equal distances from one other. No regularity of arrangement exists in the direction of the particle length; the oriented regions appear to be liquid crystalline. Upon drying, the oriented arrangement of liquid solutions is retained and it is possible to obtain oriented dry specimens.

X-rays were first used to investigate the structure of TMV by BERNAL AND FAN-KUCHEN[1]. They obtained highly oriented X-ray photographs containing a very large number of distinct reflexions. By varying the inter-particle distance, they were able to show that the X-ray pattern arose not only from the regular arrangement of the virus particles in a hexagonal lattice but in addition from the presence of a complex structure within each virus particle. The repeat distance along the fibre axis was found to be 68 A, *i.e.* much shorter than the length of the particle; they concluded that each virus particle contains a large number of equivalent sub-units. In this paper we shall deal with the internal structure of TMV and shall present newly-obtained X-ray evidence indicating that the internal regularity is based on a division of the virus into crystallographically equivalent sub-units helically arranged around the longitudinal axis of the particle.

EXPERIMENTAL METHODS

The TMV was obtained in the form of purified suspensions from Dr. R. MARKHAM of the Molteno Institute. Liquid suspensions were oriented by flow in thin capillary tubes of borosilicate glass, while oriented dry suspensions were obtained by controlled evaporation in the simple evaporation cell of BERNAL AND FANKUCHEN[1]. The X-ray photographs were taken with a rotating anode tube

* This investigation was initiated while the author was a Merck Fellow of the National Research Council (U.S.A.) at the Molteno Institute, Cambridge.

References p. 19.

The title page of Watson's TMV paper.

16   J. D. WATSON   VOL. **13** (1954)

We also indicate in Table II the value of $2\pi Rr$ at which the lower order Bessel function has its first maximum*. This value, together with a value for r, allows us to predict where the first reflexion on a given layer line can occur. (Naturally it may actually fail to appear owing to phase cancellation.) Since the diameter of TMV is 150 A, r cannot assume a value greater than 75 and so the innermost reflexions cannot occur closer to

the meridian than $\dfrac{2\pi 75}{X} = \dfrac{1}{R}$. We thus include in Table II the lowest possible value

of $1/R$ for a given r. These should be compared with the observed positions which are also tabulated in Table II. The agreement can be seen to be good, as in no case is a reflexion found in a region forbidden by the Bessel function argument. In fact, the initial reflexions systematically appear slightly beyond the predicted region and may suggest that the outermost regions of each TMV particle only slightly contribute to the

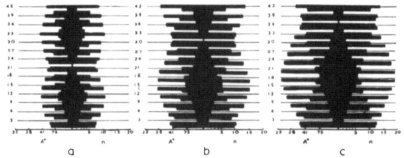

Fig. 5. Diagrammatic illustration of the forbidden regions in reciprocal space in which X-ray reflexions cannot occur. The forbidden regions are darkened and are presented for helices repeating after 3 turns and containing (a) 22, (b) 31, and (c) 37 residues per repeat of 68 A. The numbers on the left-hand margin of each diagram refer to the layer-lines along the fibre axis. The horizontal scale is described in two ways. On the left-hand side, the numbers refer to spacings in reciprocal space, while those on the right-hand side refer to the order n of the lowest order Bessel Function which contributes to a given layer line. For convenience the vertical scale has been compressed approximately six-fold and so the horizontal and vertical distances are not equivalent.

observed scattering. We should note that for very low order Bessel functions the innermost reflexions may occur very near the meridian. Thus it may be possible to observe splitting of these reflexions only under the most favourable conditions of particle orientation. It was therefore pleasing to obtain several photographs from a very well-oriented specimen which indicated a definite splitting of the 3rd and 15th order meridional reflexions. As yet we have been unable to observe splitting of the 6th, 9th or 12th meridional reflexions.

It is pertinent to compare the 31 residue helix with related helices containing 3 turns and $3n + 1$ residues per repeat. This we have done in Fig. 5, in which are illustrated diagramatically the forbidden regions of 3-turn helices containing 22, 31 and 37 residues. While all of the diagrams are roughly similar, they can be seen to differ

---

* The shape of Bessel Functions is such that they have very low values except in the immediate vicinity of maxima or minima.

References p. 19.

A page from Watson's TMV paper showing his interpretation of the calculated Bessel functions for his TMV pictures (see Chapter 13).

The following morning I anxiously awaited Francis' arrival to confirm the helical diagnosis. When he needed less than ten seconds to spot the crucial reflection, all my lingering doubts vanished. In fun I went on to trap Francis into believing that I did not think my X-ray picture was in fact very critical. Instead, I argued that the really important step was the cozy-corner insight. These flippant words were hardly out of my mouth before Francis was off on the dangers of uncritical teleology. Francis always said what he meant and assumed that I acted the same way. Though success in Cambridge conversation frequently came from saying something preposterous, hoping that someone would take you seriously, there was no need for Francis to adopt this gambit. A discourse of only one or two minutes on the emotional problems of foreign girls was always sufficient tonic for even the most staid Cambridge evening.

It was, of course, clear what we should next conquer. No more dividends could come quickly from TMV. Further unraveling of its detailed structure needed a more professional attack than I could muster. Moreover, it was not obvious that even the most backbreaking effort would give within several years the structure of the RNA component. The way to DNA was not through TMV.

The moment was thus appropriate to think seriously about some curious regularities in DNA chemistry first observed at Columbia by the Austrian-born biochemist Erwin Chargaff.[1] Since the war, Chargaff and his students had been painstakingly analyzing various

Erwin Chargaff at the 1947 Cold Spring Harbor Symposium on Quantitative Biology on Nucleic Acids and Nucleo-proteins.

DNA samples for the relative proportions of their purine and pyrimidine bases. In all their DNA preparations the number of adenine (A) molecules was very similar to the number of thymine (T) molecules, while the number of guanine (G) molecules was very close to the number of cytosine (C) molecules. More-over the proportion of adenine and thymine groups varied with their biologi-

[1] Inspired by Avery's demonstration that DNA was likely to be the heredity material, Chargaff carried out careful analyses of the chemical composition of DNA. Chargaff became a bitter critic of molecular biology, writing, for example, that "Molecular biology is essentially the practice of biochemistry without a license." His acerbic review of *The Double Helix* is reproduced in Appendix 5.

EXPERIENTIA VOL. VI/6, 1950 · pp. 201–209

BIRKHÄUSER PUBLISHERS, BASEL/SWITZERLAND

Chemical Specificity of Nucleic Acids and Mechanism of their Enzymatic Degradation[1]

By ERWIN CHARGAFF[2], New York, N.Y.

Table II [1]

Composition of desoxyribonucleic acid of ox (in moles of nitrogenous constituent per mole of P).

Constituent	Thymus			Spleen		Liver
	Prep. 1	Prep. 2	Prep. 3	Prep. 1	Prep. 2	
Adenine . .	0·26	0·28	0·30	0·25	0·26	0·26
Guanine . .	0·21	0·24	0·22	0·20	0·21	0·20
Cytosine . .	0·16	0·18	0·17	0·15	0·17	
Thymine . .	0·25	0·24	0·25	0·24	0·24	
Recovery . .	0·88	0·94	0·94	0·84	0·88	

[1] From E. CHARGAFF, E. VISCHER, R. DONIGER, C. GREEN, and F. MISANI, J. Biol. Chem. *177*, 405 (1949); and unpublished results.

Title and table from Chargaff's Experientia *paper.*

[2] A table from Chargaff's 1950 paper in *Experientia*, showing that the ratios of A:T and of G:C are approximately 1. However, Chargaff did not understand the significance of these ratios for the structure of DNA and did not, despite his later claims, discover base-pairing. He remained scathing about the double helix for many years, writing in 1962 that "DNA now is a magic name, the philosopher's stone of our days, the quintessence of contemporary alchemy."

cal origin. The DNA of some organisms had an excess of A and T, while in other forms of life there was an excess of G and C. No explanation for his striking results was offered by Chargaff, though he obviously thought they were significant. When I first reported them to Francis they did not ring a bell, and he went on thinking about other matters.[2]

Soon afterwards, however, the suspicion that the regularities were important clicked inside his head as the result of several conversations with the young theoretical chemist John Griffith. One occurred while they were drinking beer after an evening talk by the astronomer Tommy Gold on "the perfect cosmological principle."[3] Tommy's facility for making a far-out idea seem plausible set Fran-

Tommy Gold in the 1960s.

[3] Crick knew many scientists in Cambridge beyond those of his immediate circle. Watson wrote to Delbrück (December 9, 1951): "Francis has attracted around him most of the interesting young scientists in Cambridge, and so at tea in his house I am liable to meet many of the Cambridge characters, like the cosmologists Bondi, Gold, and Hoyle."

[4] The Perfect Cosmological Principle was part of the Steady State Theory, describing the nature of the universe, developed by Tommy Gold, Herman Bondi, and Fred Hoyle who were close friends. It states that the universe is uniform and unchanging in space and time. Their Steady State Theory proposed that although the universe was expanding (as required by observations of cosmological redshift), matter was continually created so that the universe appeared unchanged. Previously, George Gamow had promoted the Big Bang Theory, that the expansion of the universe arose from a single event. With the discovery of the cosmic microwave background radiation in 1965, the Big Bang has become the accepted theory.

In 1954 Gamow, with Watson, founded the RNA Tie Club, a group of scientists interested in deciphering the Genetic Code. Each member was named for an amino acid or nucleotide, had a tie pin bearing the abbreviation of their molecule's name, and wore a tie designed by Gamow and commissioned by Watson from a haberdasher in Los Angeles. Gamow is pictured wearing his tie on page 245, and a telegram to Watson from Richard Feynman, signed "Gly" for glycine, Feynman's RNA Tie Club name, is reproduced on page 248.

THE STEADY-STATE THEORY OF THE EXPANDING UNIVERSE

H. Bondi and T. Gold

(Received 1948 July 14)

Summary

The applicability of the laws of terrestrial physics to cosmology is examined critically. It is found that terrestrial physics can be used unambiguously only in a stationary homogeneous universe. Therefore a strict logical basis for cosmology exists only in such a universe. The implications of assuming these properties are investigated.

Considerations of local thermodynamics show as clearly as astronomical observations that the universe must be expanding. Hence, there must be continuous creation of matter in space at a rate which is, however, far too low for direct observation. The observable properties of such an expanding stationary homogeneous universe are obtained, and all the observational tests are found to give good agreement.

The physical properties of the creation process are considered in some detail, and the possible formulation of a field theory is critically discussed.

---

1. *The perfect cosmological principle*

1.1. The unrestricted repeatability of all experiments is the fundamental axiom of physical science. This implies that the outcome of an experiment is not affected by the position and the time at which it is carried out. A system of cosmology must be principally concerned with this fundamental assumption and, in turn, a suitable cosmology is required for its justification. In laboratory physics we have become accustomed to distinguish between conditions which can be varied at will and the inherent laws which are immutable.

Bondi and Gold's paper published in Monthly Notices of the Royal Astronomical Society *108: 252–270, 1948.*

cis to wondering whether an argument could be made for a "perfect biological principle."[4] Knowing that Griffith was interested in theoretical schemes for gene replication, he popped out with the idea that the perfect biological principle was the self-replication of the gene—that is, the ability of a gene to be exactly copied when the chromosome number doubles during cell division. Griffith, however, did not go along, since for some months he had preferred a scheme where gene copying was based upon the alternative formation of complementary surfaces.

This was not an original hypothesis. It had been floating about for almost thirty years in the circle of theoretically inclined geneticists intrigued by gene duplication. The argument went that gene duplication required the formation of a complementary (negative) image where shape was related to the original (positive) surface like a lock to a key. The complementary negative image would then function as the mold (template) for the synthesis of a new positive image. A smaller number of geneticists, however, balked at complementary replication. Prominent among them was H. J. Muller, who was impressed that several well-known theoretical physicists, especially Pascual Jordan, thought forces existed by which like attracted like.[5] But Pauling abhorred this direct mechanism and was especially irritated by the suggestion that it was supported by quantum mechanics. Just before the war, he asked Delbrück (who had drawn his attention to Jordan's papers) to coauthor a note to *Science* firmly stating that quantum mechanics favored a gene-duplicating mechanism involving the synthesis of complementary replicas.[6]

Neither Francis nor Griffith was long satisfied that evening by restatements of well-worn hypotheses. Both knew that the important task was now to pinpoint the attractive forces. Here Francis forcefully argued that specific

Pascual Jordan.

[5] **Pascual Jordan, the theoretical physicist, co-authored important papers on quantum mechanics with Werner Heisenberg and Max Born. He was keen to apply the findings of the new physics to biology: "The powerful deepening which physical knowledge has experienced since 1900...demands that its consequences be pursued beyond the domain of physics into that of the sciences of organic life." Jordan joined the Nazi party and became a SA-Mann ("Brown Shirt"). Following the war, Max Born declined to send Jordan a testimonial to aid in Jordan's de-nazification process, sending instead a list of Born's relatives who had been killed by the Nazis.**

[6] **On the left is the article in *Science* in which Pauling and Delbrück reject Pascual Jordan's arguments for attractions between identical molecules. Instead they "...feel that complementariness should be given primary consideration in the discussion of the specific attraction between molecules."**

> ### THE NATURE OF THE INTERMOLECULAR FORCES OPERATIVE IN BIOLOGICAL PROCESSES
>
> In recent papers P. Jordan[1] has advanced the idea that there exists a quantum-mechanical stabilizing interaction, operating preferentially between identical or nearly identical molecules or parts of molecules, which is of great importance for biological processes; in particular, he has suggested that this interaction might be able to influence the process of biological molecular synthesis in such a way that replicas of molecules present in the cell are formed. He has used the idea in connection with suggested explanations of the reproduction of genes, the growth of bacteriophage, the formation of antibodies, and other biological phenomena. The novelty in Jordan's work lies in his suggestion that the well-known quantum-mechanical resonance phenomenon would lead to attraction be-
>
> [1] P. Jordan, *Phys. Zeits.*, 39: 711, 1938; *Zeits. f. Phys.*, 113: 431, 1939; *Fundam. Radiol.*, 5: 43, 1939; *Zeits. f. Immun. forsch. u. exp. Ther.*, 97: 330, 1940.

Pauling and Delbrück's paper on molecular complementarity (Science 92: 77–79, 1940).

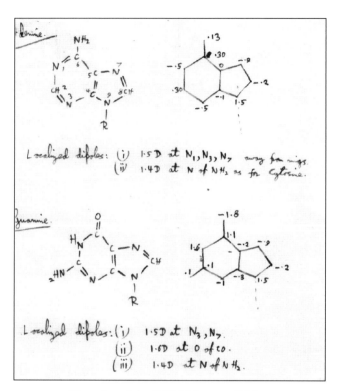

Figure from John Griffith's letter to Crick.

[7] We do not have any documentation of the discussion between Crick and Griffith, but two years later, Crick asked Griffith to carry out a similar set of calculations. Griffith's response included this page giving his estimates of the dipole forces for atoms in adenine and guanine. (March 2, 1953)

hydrogen bonds were not the answer. They could not provide the necessary exact specificity, since our chemist friends repeatedly told us that the hydrogen atoms in the purine and pyrimidine bases did not have fixed locations but randomly moved from one spot to another. Instead, Francis had the feeling that DNA replication involved specific attractive forces between the flat surfaces of the bases.

Luckily, this was the sort of force that Griffith might just be able to calculate. If the complementary scheme was right, he might find attractive forces between bases with different structures. On the other hand, if direct copying existed, his calculations might reveal attraction between identical bases. Thus, at the closing hour they parted with the understanding that Griffith would see if the calculations were feasible. Several days later, when they bumped into each other in the Cavendish tea queue, Francis learned that a semi rigorous argument hinted that adenine and thymine should stick to each other by their flat surfaces. A similar argument could be put forward for attractive forces between guanine and cytosine.[7]

Francis immediately jumped at the answer. If his memory served him right, these were the pairs of bases that Chargaff had shown to occur in equal amounts. Excitedly he told Griffith that I had recently muttered to him some odd results of Chargaff's. At the moment, though, he wasn't sure that the same base pairs were involved. But as soon as the data were checked he would drop by Griffith's rooms to set him straight.

At lunch I confirmed that Francis had got Chargaff's results right. But by then he was only routinely enthusiastic as he went over Griffith's quantum mechanical arguments. For one thing, Griffith, when pressed, did not want to defend his exact reasoning too strongly. Too many variables had been ignored to make the calculations possible in a reasonable time. Moreover, though each base has two flat sides, no explanation existed for why only one side would be chosen. And there was no reason for ruling out the idea that Chargaff's regularities had their origin in the genetic code. In some way specific groups of nucleotides must code for specific amino acids. Conceivably, adenine equaled thymine because of a yet undiscovered role in the ordering of the bases. There was in addition Roy Markham's assurance that, if Chargaff said that guanine equaled cytosine, he was equally certain it did not. In Markham's eyes, Chargaff's experimental methods inevitably underestimated the true amount of cytosine.

Nonetheless, Francis was not yet ready to dump Griffith's scheme when, early in July, John Kendrew walked into our newly acquired office to tell us that Chargaff himself would soon be in Cambridge for an evening. John had arranged for him to have dinner at Peterhouse, and Francis and I were invited to join them later for drinks in John's room. At High Table John kept the conversation away from serious matters, letting loose only the possibility

Crick doing an experiment in the Cavendish in the early 1950s.

that Francis and I were going to solve the DNA structure by model building. Chargaff, as one of the world's experts on DNA, was at first not amused by dark horses trying to win the race. Only when John reassured him by mentioning that I was not a typical American did he realize that he was about to listen to a nut. Seeing me quickly reinforced his intuition. Immediately he derided my hair and accent, for since I came from Chicago I had no right to act otherwise. Blandly telling him that I kept my hair long to avoid confusion with American Air Force personnel proved my mental instability.

The high point in Chargaff's scorn came when he led Francis into admitting that he did not remember the chemical differences among the four bases. The faux pas slipped out when Francis mentioned Griffith's calculations. Not remembering which of the bases had amino groups, he could not qualitatively describe the quantum-mechanical argument until he asked Chargaff to write out their formulas. Francis' subsequent retort that he could always look them up got nowhere in persuading Chargaff that we knew where we were going or how to get there.[8]

But regardless of what went through Chargaff's sarcastic mind, someone had to explain his results. Thus the next afternoon Francis buzzed over to Griffith's rooms in Trinity to set himself straight about the base-pair data. Hearing "Come in," he opened the door to see Griffith and a girl. Realizing that this was not the moment for science, he slowly retreated, asking Griffith to tell him again the pairs produced by his calculations. After scribbling them down on the back of an envelope, he left. Since I had departed that morning for the continent, his next stop was the Philosophical Library, where he could remove his lingering doubts about Chargaff's data. Then with both sets of information firmly in hand, he considered returning the next day to Griffith's rooms. But on second thought he realized that Griffith's interests were elsewhere. It was all too clear that the presence of popsies does not inevitably lead to a scientific future.

[8] Chargaff wrote a vivid account of this meeting. Of Crick, "…the looks of a fading racing tout…an incessant falsetto, with occasional nuggets glittering in the turbid stream of prattle." Of Watson, "…quite undeveloped at twenty-three, a grin, more sly than sheepish, saying little, nothing of consequence." Chargaff told Horace Judson that Watson and Crick "…talked so much about 'pitch' that I remember I wrote down afterwards, 'Two pitchmen in search of a helix'."

Chapter 19

[1] In 1974, Chargaff reflected on this incident. "Unfortunately retaining, as I do, only the trivialities of my past, I remember the incident at the Biochemistry Congress in 1952 and the gawky young figure, so reminiscent of one of the apprentice cobblers out of Nestroy's *Lumpazivagabundus*. I felt far from sardonic: I was looking for a toilet; but whatever door I opened, there was a lecture room and the same large portrait of Cardinal Richelieu."

The example of the meeting's elaborately designed metal name badge shown on the right belonged to Waldo Cohn, another attendee and yet another former physicist who turned to biology after his wartime involvement in creating the atomic bomb. He developed ion exchange chromatography for the separation of nucleotides and was later described by Watson as "…the only decent nucleic acid chemist in the States." (Letter to Delbrück, January 4, 1954)

Two weeks later Chargaff and I glanced at each other in Paris. Both of us were there for the International Biochemical Congress. A trace of a sardonic smile was all the recognition I got when we passed in the courtyard outside the massive Salle Richelieu of the Sorbonne.[1] That day I was tracking down Max Delbrück. Before I had left Copenhagen for Cambridge, he had offered me a research position in the biology division of Cal Tech and arranged a Polio Foundation fellowship, to start in September 1952. This March, however, I had written Delbrück that I wanted another Cambridge year. Without any hesitation he saw to it that my forthcoming fellowship was transferred to the Cavendish. Delbrück's speedy approval pleased me, for he had ambivalent feelings about the ultimate value to biology of Pauling-like structural studies.

Meeting badge for the 2eme Congrès International de Biochimie, Paris, July 1952.

Boris and Harriet Ephrussi at Cold Spring Harbor.

With the helical TMV picture now in my pocket, I felt more confident that Delbrück would at last wholeheartedly approve my liking for Cambridge. A few minutes' conversation, nonetheless, revealed no basic change in his outlook. Almost no comments emerged from Delbrück as I outlined how TMV was put together. The same indifferent response accompanied my hurriedly delivered summary of our attempts to get DNA by model building. Delbrück was drawn out only by my remark that Francis was exceedingly bright. Unfortunately, I went on to liken Francis' way of thinking to Pauling's. But in Delbrück's world no chemical thought matched the power of a genetic cross. Later that evening, when the geneticist Boris Ephrussi brought up my love affair with Cambridge, Delbrück threw up his hands in disgust.[2]

The sensation of the meeting was the unexpected appearance of Linus. Possibly because there had been considerable newspaper play on the withdrawal of his passport, the State Department reversed itself and allowed Linus to show off the α-helix.[3] A lecture was hastily arranged for the session at which Perutz spoke. Despite the short notice, an overflow crowd was on hand, hoping that they would be the first to learn of a new inspira-

[2] Born in Moscow in 1901, Boris Ephrussi took up research on fruit fly genetics with T. H. Morgan in Pasadena. He collaborated with George Beadle and their work on the genetics of eye pigmentation provided the beginnings of the "one gene, one enzyme" idea, which led to Beadle's Nobel Prize–winning work on *Neurospora*. Boris was a co-author with Watson of a spoof letter to *Nature* (see Chapter 20). Harriet had trained with Oswald Avery and was an eminent bacterial geneticist. She and Watson collaborated (1952–1953) on studies of the physical characteristics of transforming factor.

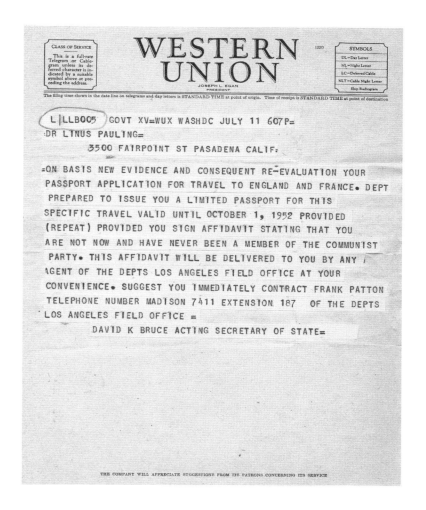

WESTERN UNION

CLASS OF SERVICE
This is a full-rate Telegram or Cablegram unless its deferred character is indicated by a suitable symbol above or preceding the address.

JOSEPH L. EGAN
PRESIDENT

1220

SYMBOLS
DL = Day Letter
NL = Night Letter
LC = Deferred Cable
NLT = Cable Night Letter
Ship Radiogram

The filing time shown in the date line on telegrams and day letters is STANDARD TIME at point of origin. Time of receipt is STANDARD TIME at point of destination

L LLB005 GOVT XV=WUX WASHDC JULY 11 607P=

DR LINUS PAULING=

3500 FAIRPOINT ST PASADENA CALIF=

=ON BASIS NEW EVIDENCE AND CONSEQUENT RE-EVALUATION YOUR
PASSPORT APPLICATION FOR TRAVEL TO ENGLAND AND FRANCE. DEPT
PREPARED TO ISSUE YOU A LIMITED PASSPORT FOR THIS
SPECIFIC TRAVEL VALID UNTIL OCTOBER 1, 1952 PROVIDED
(REPEAT) PROVIDED YOU SIGN AFFIDAVIT STATING THAT YOU
ARE NOT NOW AND HAVE NEVER BEEN A MEMBER OF THE COMMUNIST
PARTY. THIS AFFIDAVIT WILL BE DELIVERED TO YOU BY ANY
AGENT OF THE DEPTS LOS ANGELES FIELD OFFICE AT YOUR
CONVENIENCE. SUGGEST YOU IMMEDIATELY CONTRACT FRANK PATTON
TELEPHONE NUMBER MADISON 7411 EXTENSION 187 OF THE DEPTS
LOS ANGELES FIELD OFFICE =

DAVID K BRUCE ACTING SECRETARY OF STATE=

THE COMPANY WILL APPRECIATE SUGGESTIONS FROM ITS PATRONS CONCERNING ITS SERVICE

[3] After being denied a passport in May (see Chapter 17), Pauling was granted one two months later. This reversal may have been a result of the press coverage surrounding the earlier decision, but as can be seen from this telegram, it also depended on his signing an affidavit stating he had never been a communist.

tion. Pauling's talk, however, was only a humorous rehash of published ideas. It nonetheless satisfied everybody, except the few of us who knew his recent papers backward and forward. No new fireworks went off, nor was there any indication given about what now occupied his mind. After his lecture, swarms of admirers surrounded him, and I didn't have the courage to break in before he and his wife, Ava Helen, went back to the nearby Trianon Hotel.

[4] Wilkins describes the reason for his trip to Brazil in his autobiography: "...I received an invitation to go with a group of British Biomolecular scientists to visit Brazil. The plan was to visit laboratories, hold a conference on important biomolecular advances and generally liven up Brazilian science." But in the event it seems it was Brazil that rather livened up Maurice: "Rio lived up to my expectations. I was carried away by the dancing, the movements of beauty, exuberance and drama, and the joy and freedom of music."

Maurice was about, looking somewhat sour. He had stopped over on his way to Brazil, where he was to lecture for a month on biophysics.[4] His presence surprised me, since it was against his character to seek the trauma of watching two thousand bread-and-butter biochemists pile in and out of badly lighted baroque lecture halls. Speaking down to the cobblestones, he asked me whether I found the talks as tedious as he did. A few academics like Jacques Monod and Sol Spiegelman were enthusiastic speakers, but generally there was so much droning that he found it hard to stay alert for the new facts he should pick up.

I tried to rescue Maurice's morale by bringing him out to the Abbaye at Royaumont for the week-long meeting on phage following the biochemical congress.[5] Though his departure for Rio would limit him to only a night's stay, he liked the idea of meeting the people who did clever biological experiments about DNA. In the train going to Royaumont, however, he looked

[5] Watson wrote to the Cricks (August 11, 1952): "The Abbey itself is very lovely, very much like a Cambridge college and thus quite a proper atmosphere for serious talk unlike Paris with its numerous distractions...Lwoff managed to keep the meeting on a very cultured level...I'm afraid I created the only nonserious note by wearing shorts and my shirt outside on all occasions. Despite my recent haircut most of my friends thought my hair very long, and as a practical joke succeeded in putting a large amount of Visconti's perfumed (highly!) brilliantine in my hair."

An aerial view of the Abbaye at Royaumont.

The meeting at Royaumont, July 1952. Watson sits, in shorts, on the ground, front row, third from the right.

off-color, giving no indication of wanting either to read *The Times* or to hear me gossip about the phage group. After we were fixed with beds in the high-ceilinged rooms of the partially restored Cistercian monastery, I began talking with some friends I had not seen since leaving the States. Later I kept expecting Maurice to search me out, and when he missed dinner I went up to his room. There I found him lying flat on his stomach, hiding his face from the dim light I had turned on. Something eaten in Paris had not gone down properly, but he told me not to be bothered. The following morning I was given a note saying that he had recovered but had to catch the early train to Paris and apologizing for the trouble he had given me.

Later that morning Lwoff mentioned that Pauling was coming out for a few hours the next day. Immediately I began to think of ways that would allow me to sit next to him at lunch. His visit, however, bore no relation to science. Jeffries Wyman, our scientific attaché in Paris[6] and an acquaintance of Pauling's, thought that Linus and Ava Helen would enjoy the austere

Jeffries Wyman.

[6] Born into a family of Bostonian bluebloods, Wyman had a varied career as researcher and diplomat. In the latter capacity, as well as serving as scientific attaché at the U.S. Embassy in Paris, he ran the UNESCO office in Cairo. After the death of his first wife (a Cabot) he traveled widely, even living for a period with Japanese and Alaskan Eskimos, whose daily life he recorded in diaries and watercolors.

charm of the thirteenth-century buildings. During a break in the morning session I caught sight of Wyman's bony, aristocratic face in search of André Lwoff. The Paulings were here and soon began talking to the Delbrücks. Briefly I had Linus to myself after Delbrück mentioned that twelve months hence I was coming to Cal Tech. Our conversation centered on the possibility that at Pasadena I might continue X-ray work with viruses. Virtually no words went to DNA. When I brought up the X-ray pictures at King's, Linus gave the opinion that very accurate X-ray work of the type done by his associates on amino acids was vital to our eventual understanding of the nucleic acids.

I got much further with Ava Helen. Learning that I would be in Cambridge next year, she talked about her son Peter. Already I knew that Peter was accepted by Bragg to work toward a Ph.D. with John Kendrew. This was despite the fact that his Cal Tech grades left much to be desired, even considering his long bout with mononucleosis. John, however, did not want to challenge Linus' desire to place Peter with him, especially knowing that he and his beautiful blonde sister gave smashing parties. Peter and Linda, if

[7] Watson reported to the Cricks that he had given Ava Helen advice about Peter Pauling (August 11, 1952): "From Mrs. Pauling's 'small talk' I gather that Peter has not yet calmed down and so we shall not have another quiet youth like Green in our midst. I recommended to his mother a very small sum of money for him to live on and in this way may cause a tendency for the puritanical life which I am now escaping from."

Pauling family portrait including Linus, Peter, Crellin, Linda, and Ava (1947). Missing is Linus, Jr., the eldest son.

she were to visit him, would undoubtedly liven up the Cambridge scene. Then the dream of virtually every Cal Tech chemistry student was that Linda would make his reputation by marrying him. The scuttlebutt about Peter centered on girls and was confused. But now Ava Helen gave me the dope that Peter was an exceptionally fine boy whom everybody would enjoy having around as much as she did. All the same, I remained silently unconvinced that Peter would add as much to our lab as Linda. When Linus beckoned that they must go, I told Ava Helen that I would help her son adjust to the restricted life of the Cambridge research student.[7]

A garden party at Sans Souci, the country home of the Baroness Edmond de Rothschild, effectively brought the meeting to its end.[8] Dressing was no easy matter for me. Just before the biochemical congress all my belongings were snatched from my train compartment as I was sleeping. Except for a few items picked up at an army PX, the clothes I still possessed had been chosen for a subsequent visit to the Italian Alps.[9] While I felt at ease giving my talk on TMV in shorts, the French contingent feared that I would

[8] The hostess was actually the Baroness *Edouard* de Rothschild. Her husband, Baron Edouard de Rothschild, who had died 3 years earlier, built Sans Souci at Gouvieux-Chantilly, not far from Royaumont, soon after his father's death in 1906. Sans Souci bordered the runs used by the Rothschild jockeys and trainers; horse racing was a passion shared by both father and son. After her husband's death, the Baroness visited the house often until her own death in 1975. Since then it has become a culinary institute.

Sans Souci, the venue for the garden party.

[9] As indicated by this postcard to his sister, Watson undertook an extended trip that summer, not returning to Cambridge until September. As well as walking with Kendrew and Bertani, he also visited the Ephrussis and saw the Delbrücks while traveling in France, Italy, and Switzerland. See previously unpublished chapter, Appendix 2.

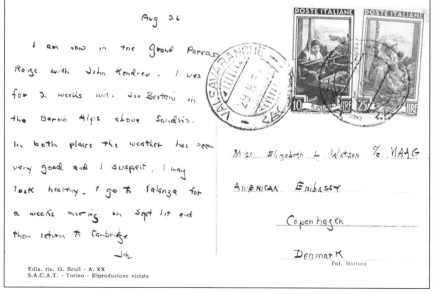

Postcard from Watson to his sister, August 26, 1952.

go one step further by arriving at Sans Souci in the same outfit. A borrowed jacket and tie, however, made me superficially presentable as our bus driver let us out in front of the huge country house.

Sol Spiegelman and I went straight for a butler carrying smoked salmon and champagne, and after a few minutes sensed the value of a cultivated aristocracy.[10] Just before we were to reboard the bus, I wandered into the large drawing room dominated by a Hals and a Rubens. The Baroness was telling several visitors how pleased she was to have such distinguished guests. She did regret, however, that the mad Englishman from Cambridge had decided not to come and enliven the mood. For an instant I was puzzled, until I realized that Lwoff had thought it prudent to warn the Baroness about an unclothed guest who might prove eccentric. The message of my first meeting with the aristocracy was clear. I would not be invited back if I acted like everyone else.

Sol Spiegelman, 1950s.

[10] At this time, Spiegelman was working on enzyme induction in bacteria and yeast and later turned to studies of RNA viruses and the replication of RNA, isolating the first RNA replicase. Spiegelman is best known for his development with Ben Hall of the technique of RNA–DNA hybridization, first in solution and later, with David Gillespie, on nitrocellulose filters.

Watson on vacation in the Italian Alps, August 1952.

Chapter 20

To Francis' dismay, I showed little tendency to concentrate on DNA when my summer holiday ended. I was preoccupied with sex, but not of a type that needed encouragement. The mating habits of bacteria were admittedly a unique conversation piece—absolutely no one in his and Odile's social circle would guess bacteria had sex lives. On the other hand, working out how they did it was best left to minor minds. Rumors of male and female bacteria were floating about at Royaumont, but not until early in September, when I attended a small meeting on microbial genetics at Pallanza, did I get the facts from the horse's mouth. There, Cavalli-Sforza and Bill Hayes talked about the experiments by which they and Joshua Lederberg had just established the existence of two discrete bacterial sexes.[1]

[1] Hayes studied in Ireland before serving as a British Medical Officer in India during the Second World War. He later set up a bacterial genetics lab at Hammersmith Hospital (see photo on page 154) which became an MRC Unit in 1957 and moved to Edinburgh in 1968. As well as his important work on bacterial genetics, he also wrote the definitive text on that field, *The Genetics of Bacteria and Their Viruses*, first published in 1964.

Bill Hayes giving an informal talk at Pallanza.

Joshua and Esther Lederberg at the 1951 Cold Spring Harbor Symposium on Quantitative Biology on Genes and Mutation.

[2] Norton Zinder, then a graduate student with Lederberg, described Lederberg's performance at the 1951 Cold Spring Harbor Symposium on Quantitative Biology on *Genes and Mutation*: "Lederberg gave a paper from our laboratory which I believe has won all competition for incomprehensibility. He spoke for more than six hours. Only H. J. Muller even began to follow it."

Bill's appearance was the sleeper of the three-day gathering: before his talk no one except Cavalli-Sforza knew he existed. As soon as he had finished his unassuming report, however, everyone in the audience knew that a bombshell had exploded in the world of Joshua Lederberg. In 1946 Joshua, then only twenty, burst upon the biological world by announcing that bacteria mated and showed genetic recombination. Since then he had carried out such a prodigious number of pretty experiments that virtually no one except Cavalli dared to work in the same field. Hearing Joshua give Rabelaisian nonstop talks of three to five hours made it all too clear that he was an *enfant terrible*. Moreover, there was his godlike quality of each year expanding in size, perhaps eventually to fill the universe.[2]

Despite Joshua's fabulous cranium, the genetics of bacteria became messier each year. Only Joshua took any enjoyment from the rabbinical complexity shrouding his recent papers. Occasionally I would try to plow through one, but inevitably I'd get stuck and put it aside for another day.[3] No high-power thoughts, however, were required to understand that the discovery of the two sexes might soon make the genetic analysis of bacteria straightforward. Conversations with Cavalli, nonetheless, hinted that Joshua was not

NATURE 701

miles and some of the treatments were destructive to other aquatic life as well as black-fly larvæ. In contrast with these results, during 1948–51 larvæ of *Simulium arcticum* Mall. were largely eradicated from sections of the Saskatchewan River for as long as 98 miles by single applications of DDT. The DDT was applied at rates as low as 0·09 p.p.m. for 16 min. as a 10 per cent solution in methylated naphthalene and kerosene.

Outstanding characteristics of the Saskatchewan River include its large rate of discharge (up to 120,000 cusec.), its freedom from aquatic vegetation, and the turbidity of the water during certain seasons of the year. During the tests, the suspended solids content of the water ranged as high as 551 p.p.m., and samples obtained by sedimentation from river water collected so far as 68 miles downstream from the point of application contained 0·24–2·26 μgm. of DDT per gram of solids. This material consisted mainly of clay and fine silt, and laboratory experiments showed that it would adsorb DDT from suspensions of 0·1 p.p.m. of DDT in distilled water.

A study of the feeding habits of the larvæ of *S. arcticum* showed that suspended particles in the river water, including much inorganic material, were consumed. It was also noted during the larvicide tests that the treatments produced much greater mortality of black-fly larvæ than of other aquatic insects, which normally do not feed on small particles suspended in the water. Quantitative samples of aquatic organisms collected before and after single applications of DDT indicated that, whereas black-fly larvæ were almost eliminated for distances ranging from 40 to 98 miles, populations of other aquatic insects were reduced by an average of 50 per cent in two tests and were unchanged in two others.

The results suggest that other fast-flowing rivers in which the water is turbid at the time of treatment might be treated similarly, and perhaps in certain clear-water streams and rivers, finely divided inorganic material with marked DDT adsorptive properties could be added along with the larvicide and kept in suspension.

F. J. H. Fredeen

Division of Entomology,
Science Service,
Canada Department of Agriculture,
Saskatoon, Sask.

A. P. Arnason

Saskatoon, Sask., now of Ottawa.

B. Berck

Winnipeg, Man.
Jan. 12.

[1] Fredeen, F. J. H., Arnason, A. P., Berck, B., and Rempel, J. G. (In preparation).
[2] Garnham, P. C. C., and McMahon, J. P., *Bull. Ent. Res.*, **37**, 619 (1947). Gjullin, C. M., Cope, A. B., Quisenberry, B. C., and DuChanois, F. R., *J. Econ. Ent.*, **42**, 1 (1949). Hocking, B., Twinn, C. R., and McDuffie, W. C., *Sci. Agric.*, **29**, 2 (1949). Hocking, B., *Sci. Agric.*, **30**, 12 (1950).

Terminology in Bacterial Genetics

The increasing complexity of bacterial genetics is illustrated by several recent letters in *Nature*[1]. What seems to us a rather chaotic growth in technical vocabulary has followed these experimental developments. This may result not infrequently in prolix and cavil publications, and important investigations may thus become unintelligible to the non-specialist. For example, the terms bacterial 'transformation', 'induction' and 'transduction' have all been used for describing aspects of a single phenomenon, namely, 'sexual recombination' in bacteria[2]. (Even the word 'infection' has found its way into reviews on this subject.) As a solution to this confusing situation, we would like to suggest the use of the term 'interbacterial information' to replace those above. It does not imply necessarily the transfer of material substances, and recognizes the possible future importance of cybernetics at the bacterial level.

Boris Ephrussi

Laboratoire de Génétique,
Université de Paris.

Urs Leopold

Zurich.

J. D. Watson

Clare College,
Cambridge.

J. J. Weigle

Institut de Physique,
Université de Genève.

[1] Lederberg, J., and Tatum, E. L., *Nature*, **158**, 558 (1946). Cavalli, L. L., and Heslot, H., *Nature*, **164**, 1058 (1949). Hayes, W., *Nature*, **169**, 118 (1952).
[2] Lindegren, C. C., *Zlb. Bakt.*, Abt. II, **92**, 40 (1935).

Histochemical Demonstration of Amine Oxidase in Liver

Dianzani[1] has shown that ditetrazolium can be used for demonstrating the activity of, among other enzymes, tyramine (amine) oxidase in mitochondria isolated from liver and kidney. The essential reaction here is a dehydrogenation[2], and hydrogen acceptors other than oxygen may be used in the oxidation of tyramine by amine oxidase[3]. It is therefore of interest that amine oxidase activity can be demonstrated in frozen sections of the tissue, using a tetrazolium compound as the hydrogen acceptor; even though the method is not entirely satisfactory, it shows the general distribution of the enzyme. Neotetrazolium[4] was found to be much more satisfactory than blue (di-)tetrazolium.

Frozen sections (15–20 μ thick) of guinea pig and rabbit liver were well washed in phosphate buffer for about thirty minutes to remove all endogenous substrates and then incubated with equal parts of 0·1 per cent neo-tetrazolium, 0·1 M phosphate buffer of pH 7·4 and 0·5 per cent tyramine solution, for 2–4 hr. At the end of the incubation period, the sections were washed in distilled water, fixed in 10 per cent neutral formalin and mounted in dilute glycerol. The use of very thin slices of liver instead of frozen sections for the incubation was found to be advantageous; they can then be fixed in formalin and sectioned on the freezing microtome. Control sections were incubated with (1) octyl alcohol[5] for three hours before incubation with tyramine and (2) potassium cyanide in a final concentration of 3×10^{-3} M, which inhibits other oxidases but has no inhibitory effect on amine oxidase.

Fat stains red with neo-tetrazolium and, together with the precipitated blue formazan, gives a general purple colour. Large fat globules can be seen in the liver, staining a bright red. This red colour can be eliminated by treating the sections with acetone, which dissolves away the fat, leaving the true (blue) colour of the precipitate. The acetone is removed by washing with water. In many cases the red colour is of advantage as it serves as a counter stain.

For the same period of incubation guinea pig liver showed a more dense precipitate than rabbit liver.

---

3 The rabbinical complexity of Lederberg's papers inspired Boris Ephrussi (pictured on page 138), Urs Leopold, Watson, and Jean Weigle (pictured on page 29) to publish a brief note on "Terminology in Bacterial Genetics" which, unbeknown to the editors of *Nature*, was a spoof. They proposed using "interbacterial information" to replace terms such as "transformation." The giveaway came in the last sentence. Doing as they suggested "…recognizes the possible future importance of cybernetics at the bacterial level."

Watson at Pallanza (back row, 2nd from left). Jacques Monod is in the front row, 2nd from left.

[4] As Watson wrote to his sister on October 27, 1952:

"…it appears that a theory which I postulated after my return from Pallanza has a very good chance of being correct as it has completely predicted the recent results of Cavalli…One more decisive experiment remains: if it comes out, it will be very pretty, as it will solve a 5-year-old paradox and allow quite rapid progress in the field of bacterial genetics…it would be nice to beat Joshua Lederberg (Wisconsin) to the solution of his life's (still rather short–he is about 28) work."

yet prepared to think simply. He liked the classical genetic assumption that male and female cells contribute equal amounts of genetic material, even though the resulting analysis was perversely complex. In contrast, Bill's reasoning started from the seemingly arbitrary hypothesis that only a fraction of the male chromosomal material enters the female cell. Given this assumption, further reasoning was infinitely simpler.

As soon as I returned to Cambridge, I beelined out to the library containing the journals to which Joshua had sent his recent work. To my delight I made sense of almost all the previously bewildering genetic crosses. A few matings still were inexplicable, but, even so, the vast masses of data now falling into place made me certain that we were on the right track. Particularly pleasing was the possibility that Joshua might be so stuck on his classical way of thinking that I would accomplish the unbelievable feat of beating him to the correct interpretation of his own experiments.[4]

My desire to clean up skeletons in Joshua's closet left Francis almost cold. The discovery that bacteria were divided into male and female sexes amused but did not arouse him. Almost all his summer had been spent collecting pedantic data for his thesis, and now he was in a mood to think about important facts. Frivolously worrying whether bacteria had one, two, or three chromosomes would not help us win the DNA structure. As long as I kept watch on the DNA literature, there was a chance that something might pop out of lunch or teatime conversations. But if I went back to pure biology, the advantage of our small headstart over Linus might suddenly vanish.

At this time there was still a nagging feeling in Francis' mind that Chargaff's rules were a real key. In fact, when I was away in the Alps he had spent a week trying to prove experimentally that in water solutions there were attractive forces between adenine and thymine, and between guanine and cytosine. But his efforts had yielded nothing. In addition, he was really never at ease talking with Griffith. Somehow their brains didn't jibe well, and there would be long awkward pauses after Francis had thrashed through the merits of a given hypothesis. This was no reason, however, not to tell Maurice that conceivably adenine was attracted to thymine and guanine to cytosine. Since he had to be in London late in October for another reason, he dropped a line to Maurice saying he could come by King's. The reply, inviting him to lunch, was unexpectedly cheerful, and so Francis looked forward to a realistic discussion of DNA.

However, he made the mistake of tactfully appearing not too interested in DNA by starting to talk about proteins. Over half the lunch was thus wasted when Maurice changed the topic to Rosy and droned on and on about her lack of cooperation.[5] Meantime, Francis' mind fastened on a more amusing topic until, the meal over, he remembered that he had to rush to a 2:30 appointment. Hurriedly he left the building and was out on the street before realizing he had not brought up the agreement between Griffith's calculations and Chargaff's data. Since it would look too silly to rush back in, he went on, returning that evening to Cambridge. The following morning, after I was told about the futility of the lunch, Francis tried to generate enthusiasm for our having a second go at the structure.

Another zeroing in on DNA, however, did not make sense to me. No fresh facts had come in to chase away the stale taste of last winter's debacle. The only new result we were likely to pick up before Christmas was the divalent metal content of the DNA containing phage T4. A high value, if found, would strongly suggest binding of Mg^{++} to DNA. With such evidence I might at last force the King's groups to analyze their DNA samples. But the prospects for immediate hard results were not good. First, Maaløe's colleague Niels Jerne must send the phage from Copenhagen. Then I would need to arrange for accurate measurement of both the divalent metals and the DNA content. Finally, Rosy would have to budge.[6]

Fortunately, Linus did not look like an immediate threat on the DNA front. Peter Pauling arrived with the inside news that his father was preoccupied with schemes for the supercoiling of α-helices in the hair protein, keratin. This was not especially good news to Francis. For almost a year he had been in and out of euphoric moods about how α-helices packed together in coiled coils. The trouble was that his mathematics never gelled tightly. When pressed he admitted that his argument had a woolly component. Now

[5] Wilkins later wrote to Crick (early 1952): "Franklin barks often but doesn't succeed in biting me. Since I reorganized my time so that I can concentrate on the job, she no longer gets under my skin. I was in a bad way about it all when I last saw you."

[6] Watson in a letter of May 20, 1952 described his plans to Delbrück: "I have just arranged for a government lab to do a complete cation analysis of some purified T4 which Taj Jerne is sending from Copenhagen. I hope to have the results by the time of the Royaumont meeting." Jerne is pictured on page 33.

he faced the possibility that Linus' solution would be no better and yet he would get all the credit for the coiled coils.

Experimental work for his thesis was broken off so that the coiled-coil equations could be taken up with redoubled effort. This time the correct equations fell out, partly thanks to the help of Kreisel, who had come over to Cambridge to spend a weekend with Francis. A letter to *Nature* was quickly drafted and given

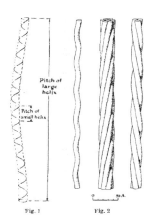

A figure from Pauling and Corey's paper on coiled coils of α-keratin. On the left, the α-helical polypeptide chain is shown to follow a larger helical course, and on the right how α-helices can pack to form three-strand and seven-strand cables.

[7] *Crick, Pauling, and coiled-coils*

Crick's work on coiled coils landed him in trouble when the editor of *Nature* hurried publication of Crick's paper and delayed Pauling's. Pauling's paper arrived at *Nature* on October 14, 1952, and while Crick sent his paper in one week later, his was published on November 22, six weeks before Pauling's paper appeared on January 10, 1953! According to Peter Pauling, in the end Crick and Kendrew "…finally got Bragg to write a letter to the editor telling him to get off the pot."

The matter was further complicated when Crick suggested that Pauling had got the idea of coiled coils from him (Crick) following a conversation they had had in the summer of 1952. Pauling got to hear the rumor and wrote to Perutz on March 29, 1953: "I am writing to try and clear up a situation that may have caused you some concern…Word has reached me that you felt that I had obtained the idea of coiling helixes, in alpha-keratin, from Crick, and had not acknowledged my indebtedness to him…Mr. Crick asked me if I had ever thought of the possibility that the alpha-helixes were twisted about one another. I answered that I had…The idea was not a new one to me then, but it had not yet been well worked out, and it seems to me that the way in which we worked it out, as described in our paper in *Nature*, is much different from that discussed by Crick. I hope that if my memory about the discussion in Cambridge is at fault Mr. Crick will let me know."

Crick replied on April 14, that "It was natural, therefore, when Peter told us you were working on coiled-coils that the idea got around that I had suggested the idea to you." But Crick confirmed that their approaches were different: "…it was clear to me there were very little grounds, if any, for such a belief. In particular you had suggested a definite model, whereas I had not, and, more important, you had put forward a different reason for coiling." However, Crick added, "On reflection I think it might have made things easier if you had let me know you were writing a paper on the idea, so that I would have had the opportunity of putting forward my ideas simultaneously. However, as things turned out, thanks to the many channels of communication between Caltech and the Cavendish, this is effectively what happened."

to Bragg to send on to the editors, with a covering note asking for speedy publication. If the editors were told that a British article was of above-average interest, they would try to publish the manuscript almost immediately. With luck, Francis' coiled coils would get into print as soon as if not before Pauling's.[7]

Thus there was growing acceptance both in and outside Cambridge that Francis' brain was a genuine asset. Though a few dissidents still thought he was a laughing talking-machine, he nonetheless saw problems through to the finish line. A reflection of his increasing stature was an offer received early in the fall to join David Harker in Brooklyn for a year. Harker, having collected a million dollars to solve the structure of the enzyme ribonuclease, was in search of talent, and the offer of six thousand for one year seemed to Odile wonderfully generous.[8] As expected, Francis had mixed feelings. There must be reasons why there were so many jokes about Brooklyn. On the other hand, he had never been in the States, and even Brooklyn would provide a base from which he could visit more agreeable regions. Also, if Bragg knew that Crick would be away for a year, he might view more favorably a request from Max and John that Francis be reappointed for another three years after his thesis was submitted. The best course seemed tentatively to accept the offer, and in mid-October he wrote Harker that he would come to Brooklyn in the fall of the following year.

As the fall progressed, I remained ensnared by bacterial matings, often going up to London to talk with Bill Hayes at his Hammersmith Hospital lab.[9] My mind snapped back to DNA on the evenings when I managed to catch Maurice for dinner on my way home to Cambridge. Some afternoons, however, he would quietly slip away, and his lab group had it that a special girl friend existed. Finally it came out that everything was above board. The afternoons were spent at a gymnasium learning how to fence.

The situation with Rosy remained as sticky as ever. Upon his return from Brazil, the unmistakable impression was given that she considered collaboration even more impossible than before. Thus, for relief, Maurice had taken up interference microscopy to find a trick for weighing chromosomes. The question of finding Rosy a job elsewhere had been brought to his boss, Ran-

[8] In 1949, Irving Langmuir asked David Harker, an X-ray crystallographer, what he would do with one million dollars. Harker's response was that he would take ten years off and use the time (and money) to determine the structure of a protein. Langmuir raised the one million dollars from foundations and in 1950, Harker created the Protein Structure Project at the Polytechnic Institute of Brooklyn.

[9] Hayes' laboratory was in the Postgraduate Medical School of London,
later known as the Royal Postgraduate
Medical School. The latter was located
on the site of the Hammersmith Hospital in West London.

Hammersmith Hospital, London.

dall, but the best to be hoped for would be a new position starting a year
hence. Sacking her immediately on the basis of her acid smile just couldn't
be done.[10] Moreover, her X-ray pictures were getting prettier and prettier.
She gave no sign, however, of liking helices any better. In addition, she
thought there was evidence that the sugar-phosphate backbone was on the
outside of the molecule. There was no easy way to judge whether this assertion had any scientific basis. As long as Francis and I remained closed
out from the experimental data, the best course was to maintain an open
mind. So I returned to my thoughts about sex.

[10] Franklin was as keen to leave King's as Wilkins was to see her go. In the box opposite we record her correspondence on the matter with her friend Anne Sayre and her
future boss, J. D. Bernal.

Franklin's move to Birkbeck

In a March 1952 letter, Franklin brought her friends Anne and David Sayre up to date on the wretched state of things at King's: "When I got back from my summer holidays I had a terrific crises with Wilkins, which nearly resulted in my going straight back to Paris. Since then we have agreed to differ, and work goes on—going on, in fact, quite well. I went back to Paris for a week in January to decide whether or not to go back. Got it all fixed up to work with Vittorio on liquids, then thought it was probably silly after all. Somehow I feel that to try to take such a big step backwards wouldn't work. So I went to see Bernal, who condescended to recognize me, made himself pleasant, and gave me some hopes of working in his biological group one day…Whatever one may have against the man he's brilliant, and I should think an inspiring person to work under."

Anne Sayre was sympathetic to Franklin's struggles at King's (from her reply of March 8: "if you murder anyone at King's, I will fly over as a character witness and swear it was justifiable homicide."). But she also had reservations about the proposed move to Birkbeck: "As you know, my anti-Bernal feelings are passionate, so I can't help looking with dismay on your plan to join his group." Nevertheless the plan moved forward and Franklin wrote again to the Sayres from Yugoslavia on June 2: "I've seen Bernal again, and he will take me any time if Randall agrees, but I decided it would be bad politics to talk to Randall just before going away for a month, so that's a pleasure in store for when I get back." Once back, she did speak to Randall, as indicated in the letter to Bernal shown here.

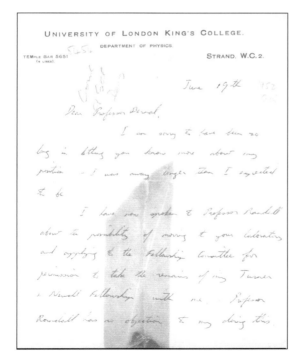

Franklin's letter to Bernal discussing her moving to Birkbeck College. The year "1952" was added by Aaron Klug ("AK").

Dear Professor Bernal,

I am sorry to have been so long in letting you know more about my position—I was away longer than I expected to be.

I have now spoken to Professor Randall about the possibility of moving to your laboratory and applying to the Fellowship Committee for permission to take the remains of my Turner and Newall Fellowship with me. Professor Randall has no objection to my doing this.

I hope this arrangement is still agreeable to you. If it is, would you suggest a time when I could come to see you to discuss things in more detail?

Yours sincerely,
Rosalind Franklin

Chapter 21

I was by now living in Clare College. Soon after my arrival at the Cavendish, Max had slipped me into Clare as a research student. Working for another Ph.D. was nonsense, but only by using this dodge would I have the possibility of college rooms. Clare was an unexpectedly happy choice. Not only was it on the Cam with a perfect garden but, as I was to learn later, it was especially considerate toward Americans.[1]

Before this happened I was almost stuck in Jesus. At short notice Max and John thought I would have the best chance to be accepted by one of the

[1] Clare College, the second oldest of Cambridge University, was founded in 1326. It is pictured in Chapter 6, and its Old Court, which runs alongside King's College chapel, is seen in the famous photo taken by Freddie Gutfreund of Watson and Crick strolling along the backs after lunch one day in October 1952 (Chapter 7).

In a letter from Watson to his sister, October 8, 1952:

"I now live in college and rather like it. My rooms are pleasantly large but rather dull. However, with Odile's help I hope to liven them up." And again on October 18: "...Clare food remains impossible and so I eat many of my meals in the English Speaking Union. I also have food in my college rooms since I find I'm quite hungry about 12 p.m. To my surprise I find I can make tea." Watson remains attached to his former college and in 2005 donated the sculpture of the double helix by Charles Jencks shown here.

Charles Jencks, Double Helix at Clare College.

[2] Denys Wilkinson was later Professor of Experimental Physics in the University of Oxford and was knighted in 1974. He designed the Wilkinson Analogue-to-Digital converter.

[3] Nicholas G. L. Hammond was a scholar of ancient Greece renowned for his work on Alexander The Great. His familiarity with Greece and Albania, and his fluency in the languages, was exploited by the Allies during the war when he was recruited as an officer in the British Special Operations Executive, involving himself in many sabotage missions. After the war he returned to academia, in the early 1950s serving as Senior Tutor at Clare College where Watson encountered him.

small colleges, since they had relatively fewer research students than the large, more prestigious, and wealthy colleges like Trinity or King's. Max thus asked the physicist Denis Wilkinson, then a fellow of Jesus, whether an opening might exist in his college. The following day Denis came by to say Jesus would have me and that I should arrange an appointment to learn the formalities of matriculation.[2]

A talk with its head tutor, however, made me try elsewhere. Jesus' possession of only a few research students appeared related to its formidable reputation on the river. No research student could live in, and so the only predictable consequences of being a Jesus man were bills for a Ph.D. that I would never acquire. Nick Hammond, the classicist head tutor at Clare, painted a much rosier outlook for their foreign research students. In my second year up, I could move into the college. Moreover, at Clare there would be several American research students I might meet.[3]

Nonetheless, during my first Cambridge year, when I lived on Tennis Court Road with the Kendrews, I saw virtually nothing of college life. After matriculation I went into hall for several meals until I discovered that I was unlikely to meet anyone during the ten-to-twelve-minute interval needed to slop down the brown soup, stringy meat, and heavy pudding provided on most evenings. Even during my second Cambridge year, when I moved into rooms on the R staircase of Clare's Memorial Court, my boycott of college food continued. Breakfast at the Whim could occur much later than if I went to hall. For 3/6 the Whim gave a half-warm site to read *The Times* while flat-capped Trinity types turned the pages of the *Telegraph* or *News Chronicle*. Finding

Watson frequented the Whim for breakfast.

Bath Hotel, a place for special occasions.

suitable evening food on the town was trickier. Eating at the Arts or the Bath Hotel was reserved for special occasions, so when Odile or Elizabeth Kendrew did not invite me to supper I took in the poison put out by the local Indian and Cypriote establishments.

My stomach lasted only until early November before violent pains hit me almost every evening. Alternative treatments with baking soda and milk did not help, and so, despite Elizabeth's assurance that nothing was wrong, I showed up at the ice-cold Trinity Street surgery of a local doctor. After I was allowed to appreciate the oars on his walls, I was expelled with a prescription for a large bottle of white fluid to be taken after meals.[4] This kept me going for almost two weeks, when, with the bottle empty, I returned to the surgery with the fear that I had an ulcer. The news that an alien's dyspeptic pains were persisting did not, however, evoke any sympathetic

The white fluid was presumably Milk of Magnesia.

[4] The local doctor Watson visited was Dr. Edward Bevan, a rowing enthusiast who coached the university team and had been a member of the British coxless fours that won a gold medal at the 1928 Olympics. Remarkably this is the same Dr. Bevan who diagnosed and treated Ludwig Wittgenstein for prostate cancer. The philosopher even moved into Bevan's home in Storey Way for the last two months of his life, dying there in April 1951, just six months before Watson arrived in Cambridge.

Drawing of 19 Portugal Place by Odile Crick, used as a Christmas card by the Cricks circa 1960.

Picture of the Cricks with Freddie Gutfreund and Christine Bennett (later Lady Jennings) in Portugal Place in the mid-1950s.

words, and again I retreated into Trinity Street with a prescription for more white stuff.

That evening I stopped by at the Cricks' newly bought house, hoping that gossip with Odile would make me forget my stomach. The Green Door recently had been abandoned for larger quarters on nearby Portugal Place. Already the dreary wallpaper on the lower floors was gone, and Odile was busy making curtains appropriate for a house large enough to have a bathroom. After I was given a glass of warm milk we began discussing Peter Pauling's discovery of Nina, Max's young Danish au pair girl. Then the problem was taken up of how I might establish a connection with the high-class boarding house run by Camille "Pop" Prior at 8 Scroope Terrace. The food at Pop's would offer no improvement over hall, but the French girls who came to Cambridge to improve their English were another matter.[5] A seat at Pop's dinner table, however, could not be asked for directly. Instead, both Odile and Francis thought the best tactic for getting a foot in the door was

[5] Writing to Delbrück, December 9, 1951, Watson bemoans the difficulty of finding lively female company in Cambridge. "As you can no doubt guess, women at Cambridge and Oxford are very rare, and so much ingenuity must be spent in finding lively and pretty girls for one's party."

[6] Watson's French lessons are frequently mentioned in his letters to his sister in the last months of 1952. On October 8 he reports that "I have started taking private French lessons from the famed Mrs. Camille Prior who runs the 'high class boarding house' for young Continental girls. They should be rather pleasant as well as instructive."

Mrs. Camille Prior was well known in Cambridge theatrical circles. Described as the "...indefatigable producer of every kind of dramatic and musical show...," she was noted particularly for the historical pageant she staged each year.

to commence French lessons with Pop, whose deceased husband had been the Professor of French before the war. If I suited Pop's fancy, I might be invited to one of her sherry parties and meet her current crop of foreign girls. Odile promised to ring Pop to see if lessons could be arranged, and I cycled back to college with the hope that soon my stomach pains would have reason to vanish.[6]

Back in my rooms I lit the coal fire, knowing there was no chance that the sight of my breath would disappear before I was ready for bed. With my fingers too cold to write legibly I huddled next to the fireplace, daydreaming about how several DNA chains could fold together in a pretty and hopefully scientific way. Soon, however, I abandoned thinking at the molecular level and turned to the much easier job of reading biochemical papers on the interrelations of DNA, RNA, and protein synthesis.

Virtually all the evidence then available made me believe that DNA was the template upon which RNA chains were made. In turn, RNA chains were the likely candidates for the templates for protein synthesis. There were some fuzzy data using sea urchins, interpreted as a transformation of DNA into RNA, but I preferred to trust other experiments showing that DNA molecules, once synthesized, are very very stable. The idea of the genes' being immortal smelled right, and so on the wall above my desk I taped up a paper sheet saying DNA → RNA → protein. The arrows did not signify chemical transformations, but instead expressed the transfer of genetic information from the sequences of nucleotides in DNA molecules to the sequences of amino acids in proteins.

Though I fell asleep contented with the thought that I understood the relationship between nucleic acids and protein synthesis, the chill of dressing in an ice-cold bedroom brought me back to the knowing truth that a slogan was no substitute for the DNA structure. Without it, the only impact that Francis and I were likely to have was to convince the biochemists we met in a nearby pub that we would never appreciate the fundamental significance of complexity in biology. What was worse, even when Francis stopped thinking about coiled coils or I about bacterial genetics, we still remained stuck at the same place we were twelve months before. Lunches at the Eagle frequently went by without a mention of DNA, though usually somewhere on

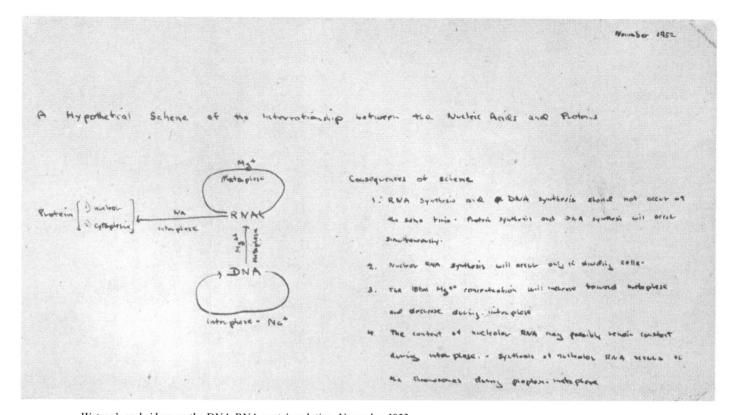

Watson's early ideas on the DNA-RNA-protein relation, November 1952.

our after-lunch walk along the backs genes would creep in for a moment.

On a few walks our enthusiasm would build up to the point that we fiddled with the models when we got back to our office. But almost immediately Francis saw that the reasoning which had momentarily given us hope led nowhere. Then he would go back to the examination of the hemoglobin X-ray photographs out of which his thesis must emerge. Several times I carried on alone for a half hour or so, but without Francis' reassuring chatter my inability to think in three dimensions became all too apparent.

I was thus not at all displeased that we were sharing our office with Peter Pauling, then living in the Peterhouse hostel as a research student of John

[7] Pauling began thinking intensively about the structure of DNA in November 1952. These are two pages from his notebook.

On the first page (page 1) he exclaims, "Perhaps we have a triple-chain structure!" The diagram on the second (page 13) resembles that in the published paper (Chapter 22).

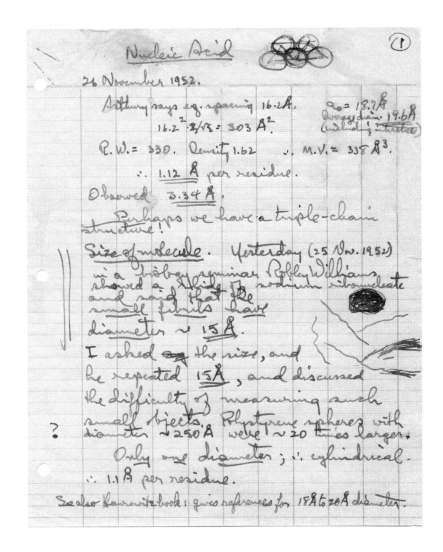

Kendrew's. Peter's presence meant that, whenever more science was pointless, the conversation could dwell on the comparative virtues of girls from England, the Continent, and California. A fetching face, however, had nothing to do with the broad grin on Peter's face when he sauntered into the office one afternoon in the middle of December and put his feet up on his desk.

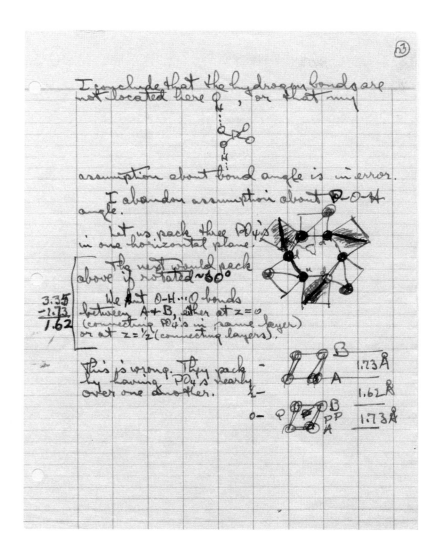

Peter Pauling.

[8] In reply to this letter from his father, Peter wrote on January 13 saying that he would like a copy.

"I would appreciate very much a copy of 'your' article. The MRC Unit would like one too. They are very interested. A couple of months ago Corey wrote Donahue or someone and said ' "We" are writing another article…' With the we in quotation marks. I don't know if I should tell you all this. It is pretty amusing."

In his hand was a letter from the States that he had picked up on his return to Peterhouse for lunch.

It was from his father. In addition to routine family gossip was the long-feared news that Linus now had a structure for DNA.[7,8] No details were given of what he was up to, and so each time the letter passed between Fran-

[9] In addition to his son, Pauling also wrote to Randall on December 31 letting him know there was a structure, but giving no details.

"Professor Corey and I are especially happy during this holiday season. We have been attacking the problem of the structure of nucleic acid during recent months, and have discovered a structure which we think may be the structure of the nucleic acids—that is, we feel that the nucleic acid molecule may have one and only one stable structure. Our first paper on this subject has been submitted for publication."

The letter from Pauling to Randall also mentions that Pauling's lab are taking their own X-ray diffraction photos. These were being done by Alex Rich, pictured opposite a year or two later in Portugal Place with Odile Crick and Hugh Huxley.

31 December 1952

Professor J. T. Randall, F.R.S.
King's College
University of London
Strand, W. C. 2
London, England

Dear Randall:

I am pleased to have your letter of 23 December, and to learn about the meeting on the Nature and Structure of Connective Tissues, that you are organizing for 26 and 27 March.

I am not sure at the present time whether any of the members of our staff will be able to attend the meeting. I myself am planning to come to Europe for the Ninth Solvay Congress, 8-14 April. This trip, however, requires me to be absent from the California Institute of Technology during term time, and I am accordingly planning to make it as short a trip as possible. It seems unlikely that I shall be able to leave Pasadena nearly two weeks earlier than would otherwise be necessary, in order to participate in your meeting, although, of course, I am very much interested in the subject.

Professor Corey and I are especially happy during this holiday season. We have been attacking the problem of the structure of nucleic acid during recent months, and have discovered a structure which we think may be the structure of the nucleic acids — that is, we feel that the nucleic acid molecule may have one and only one stable structure. Our first paper on this subject has been submitted for publication. I regret to say that our x-ray photographs of sodium thymonucleate are not especially good; I have never seen the photographs made in your laboratory, but I understand that they are much better than those of Astbury and Bell, whereas ours are inferior to Astbury and Bell's. We are hoping to obtain better photographs, but fortunately the photographs that we have are good enough to permit the derivation of our structure.

Sincerely yours,

Linus Pauling:W

Linus Pauling's letter to Randall about his paper.

Photo of Hugh Huxley, Alex Rich, and Odile Crick in Portugal Place.

cis and me the greater was our frustration. Francis then began pacing up and down the room thinking aloud, hoping that in a great intellectual fervor he could reconstruct what Linus might have done. As long as Linus had not told us the answer, we should get equal credit if we announced it at the same time.

Nothing worthwhile had emerged, though, by the time we walked up-stairs to tea and told Max and John of the letter. Bragg was in for a moment, but neither of us wanted the perverse joy of informing him that the English labs were again about to be humiliated by the Americans. As we munched chocolate biscuits, John tried to cheer us up with the possibility of Linus' being wrong. After all, he had never seen Maurice's and Rosy's pictures. Our hearts, however, told us otherwise.[9]

Chapter 22

No further news emerged from Pasadena before Christmas.[1] Our spirits slowly went up, for if Pauling had found a really exciting answer the secret could not be kept long. One of his graduate students must certainly know what his model looked like, and if there were obvious biological implications the rumor would have quickly reached us. Even if Linus was somewhere near the right structure, the odds seemed against his getting near the secret of gene replication. Also, the more we thought about DNA chemistry, the more unlikely seemed the possibility that even Linus could pick off the structure in total ignorance of the work at King's.

Maurice was told that Pauling was in his pasture when I passed through London on my way to Switzerland for a Christmas skiing holiday. I was hoping that the urgency created by Linus' assault on DNA might make him ask

[1] Pasadena, the home of the California Institute of Technology. Caltech and the Department of Chemistry were Pauling's professional home from 1927 to 1964.

The Crellin Laboratory of Chemistry, California Institute of Technology, Pasadena, 1940s.

Francis and me for help. However, if Maurice thought that Linus had a chance to steal the prize, he didn't let on. Much more important was the news that Rosy's days at King's were numbered. She had told Maurice that she wanted soon to transfer to Bernal's lab at Birkbeck College.[2] Moreover, to Maurice's surprise and relief, she would not take the DNA problem with her. In the next several months she was to conclude her stay by writing up her work for publication. Then, with Rosy at last out of his life, he would commence an all-out search for the structure.[3]

Upon my return to Cambridge in mid-January, I sought out Peter to learn what was in his recent letters from home. Except for one brief reference to DNA, all the news was family gossip. The one pertinent item, however, was not reassuring. A manuscript on DNA had been written, a copy of which would

[2] Writing of her imminent move, Franklin told her friend Adrienne Weill (March 10, 1953): "I start at Birkbeck next week. It got delayed repeatedly—first because I missed a month with flu and things in the autumn, then because I thought another month would give me a lot more results—but it didn't, and I'm abandoning an unfinished job now in order to get out of King's without further delay…As far as the lab is concerned, I shall be moving from a palace to a slum, but I'm sure I shall find Birkbeck pleasanter all the same." Franklin was indeed much happier at Birkbeck, though some of Anne Sayre's concerns were valid. In December 1953, she wrote to the Sayres, alluding to Bernal's communist sympathies: "Birkbeck is an improvement on King's, as it couldn't fail to be. But the disadvantages of Bernal's group are obvious—a lot of narrow-mindedness, and obstruction directed especially against those who are not Party members."

[3] Randall reminded Franklin that she was not to work on DNA in a letter of April 17, 1953, after she had moved to Birkbeck. While it is not unusual for a scientist to give up a project when moving to a new laboratory, the peremptory tone of Randall's letter is far from collegial.

UNIVERSITY OF LONDON KING'S COLLEGE.

From The Wheatstone Professor of Physics.
J. T. RANDALL, F.R.S.

TEMPLE BAR 5653.   STRAND, W.C.2.

Miss R.E. Franklin,
Birkbeck College Research Laboratory,
21 Torrington Square,
London, W.C.1
                                   17th April 1953

Dear Miss Franklin,

        You will no doubt remember that when we discussed the question of your leaving my laboratory you agreed that it would be better for you to cease to work on the nucleic acid problem and take up something else. I appreciate that it is difficult to stop thinking immediately about a subject on which you have been so deeply engaged, but I should be grateful if you could now clear up, or write up, the work to the appropriate stage. A very real point about which I am a little troubled is that it is obviously not right that Gosling should be supervised by someone not specifically resident in this laboratory. You will realise that the necessary reorganisation for this purpose which arises from your departure cannot really proceed while you remain, in an intellectual sense, a member of the laboratory.

        Yours sincerely,

        JTRandall

soon be sent to Peter. Again there was not a hint of what the model looked like. While waiting for the manuscript to arrive, I kept my nerves in check by writing up my ideas on bacterial sexuality. A quick visit to Cavalli in Milan, which occurred just after my skiing holiday in Zermatt, had convinced me that my speculations about how bacteria mated were likely to be right. Since I was afraid that Lederberg might soon see the same light, I was anxious to publish quickly a joint article with Bill Hayes. But this manuscript was not in final form when, in the first week of February, the Pauling paper crossed the Atlantic.

Two copies, in fact, were dispatched to Cambridge—one to Sir Lawrence, the other to Peter. Bragg's response upon receiving it was to put it aside. Not knowing that Peter would also get a copy, he hesitated to take the manuscript down to Max's office. There Francis would see it and set off on another wild-goose chase. Under the present timetable there were only eight months more of Francis' laugh to bear. That is, if his thesis was finished on schedule. Then for a year, if not more, with Crick in exile in Brooklyn, peace and serenity would prevail.

GENETIC EXCHANGE IN ESCHERICHIA COLI K12: EVIDENCE FOR THREE LINKAGE GROUPS

By J. D. Watson* and W. Hayes

Medical Research Council Unit for the Study of the Molecular Structure of Biological Systems, The Cavendish Laboratory, Cambridge, England; and Department of Bacteriology, Postgraduate Medical School, London, England

Communicated by M. Delbrück, February 27, 1953

The genetic analysis of recombination within strain K-12 of *Escherichia coli* has, until recently, been considered entirely in terms of the hypothesis of an orthodox sexual mechanism involving union of entire cells, followed by a normal meiotic cycle of chromosome pairing, crossing over, and reduction.[1] The evidence in favor of this assumption was many-sided and included both the existence of unstable strains which behaved like heterozygous diploids by segregating out stable recombinant strains, and the inability of cell-free filtrates (in contrast to type transformation in pneumococcus) to induce recombination. Since recombination is a rare event, it has not been possible to observe zygote formation and the evidence for a classical genetic mechanism has remained circumstantial.

The Watson and Hayes paper on bacterial sexuality was published in the Proceedings of the National Academy of Sciences, *May 1953. This paper and its publication are discussed further in Chapter 26.*

While Sir Lawrence was pondering whether to chance taking Crick's mind off his thesis, Francis and I were poring over the copy that Peter brought in after lunch. Peter's face betrayed something important as he entered the door, and my stomach sank in apprehension at learning that all was lost. Seeing that neither Francis nor I could bear any further suspense, he quickly told us that the model was a three-chain helix with the sugar-phosphate backbone in the center. This sounded so suspiciously like our aborted

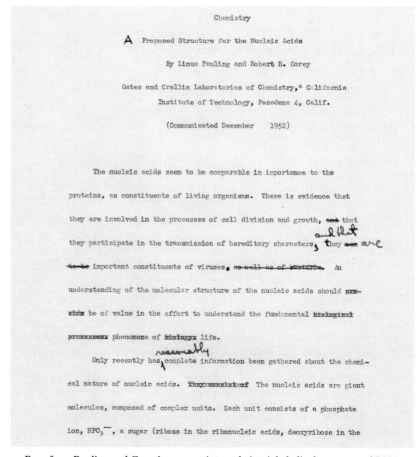

Chemistry

A Proposed Structure for the Nucleic Acids

By Linus Pauling and Robert B. Corey

Gates and Crellin Laboratories of Chemistry,* California
Institute of Technology, Pasadena 4, Calif.

(Communicated December   1952)

The nucleic acids seem to be comparable in importance to the proteins, as constituents of living organisms. There is evidence that they are involved in the processes of cell division and growth, and that they participate in the transmission of hereditary characters, they are important constituents of viruses. An understanding of the molecular structure of the nucleic acids should be of value in the effort to understand the fundamental phenomena of life.

Only recently has complete information been gathered about the chemical nature of nucleic acids. The nucleic acids are giant molecules, composed of complex units. Each unit consists of a phosphate ion, HPO_3^{--}, a sugar (ribose in the ribonucleic acids, deoxyribose in the

Page from Pauling and Corey's manuscript on their triple helical structure of DNA.

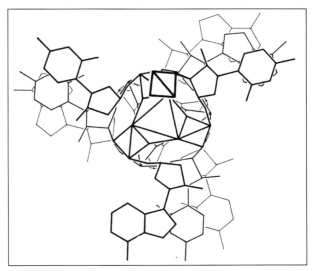

A figure from Pauling and Corey's paper "A Proposed Structure for the Nucleic Acids," Proceedings of the National Academy of Sciences, *February 1953. A view down the triple helix with three phosphate-sugar backbones and the bases on the outside, extending out from the core. (See Pauling's notebook sketch, Chapter 21.)*

effort of last year that immediately I wondered whether we might already have had the credit and glory of a great discovery if Bragg had not held us back. Giving Francis no chance to ask for the manuscript, I pulled it out of Peter's outside coat pocket and began reading. By spending less than a minute with the summary and the introduction, I was soon at the figures showing the locations of the essential atoms.

At once I felt something was not right. I could not pinpoint the mistake, however, until I looked at the illustrations for several minutes. Then I realized that the phosphate groups in Linus' model were not ionized, but that each group contained a bound hydrogen atom and so had no net charge. Pauling's nucleic acid in a sense was not an acid at all. Moreover, the uncharged phosphate groups were not incidental features. The hydrogens were part of the hydrogen bonds that held together the three intertwined chains. Without the hydrogen atoms, the chains would immediately fly apart and the structure vanish.[4]

[4] Verner Schomaker, pictured on page 225, a colleague of Pauling, is said to have commented on Pauling's triple helix: "If that were the structure of DNA, it would explode!"

Everything I knew about nucleic-acid chemistry indicated that phosphate groups never contained bound hydrogen atoms. No one had ever questioned that DNA was a moderately strong acid. Thus, under physiological conditions, there would always be positively charged ions like sodium or magnesium lying nearby to neutralize the negatively charged phosphate groups. All our speculations about whether divalent ions held the chains together would have made no sense if there were hydrogen atoms firmly bound to the phosphates. Yet somehow Linus, unquestionably the world's most astute chemist, had come to the opposite conclusion.

When Francis was amazed equally by Pauling's unorthodox chemistry, I began to breathe slower. By then I knew we were still in the game. Neither of us, however, had the slightest clue to the steps that had led Linus to his blunder. If a student had made a similar mistake, he would be thought unfit to benefit from Cal Tech's chemistry faculty. Thus, we could not but initially worry whether Linus' model followed from a revolutionary re-evaluation of the acid-base properties of very large molecules. The tone of the manuscript, however, argued against any such advance in chemical theory. No reason existed to keep secret a first-rate theoretical breakthrough. Rather, if that had occurred Linus would have written two papers, the first describing his new theory, the second showing how it was used to solve the DNA structure.

The blooper was too unbelievable to keep secret for more than a few minutes. I dashed over to Roy Markham's lab to spurt out the news and to receive further reassurance that Linus' chemistry was screwy. Markham predictably expressed pleasure that a giant had forgotten elementary college chemistry. He then could not refrain from revealing how one of Cambridge's great men had on occasion also forgotten his chemistry. Next I hopped over to the organic chemists, where again I heard the soothing words that DNA was an acid.

By teatime I was back in the Cavendish, where Francis was explaining to John and Max that no further time must be lost on this side of the Atlantic. When his mistake became known, Linus would not stop until he had captured the right structure. Now our immediate hope was that his chemical colleagues would be more than ever awed by his intellect and not probe the details of his model. But since the manuscript had already been dispatched to the *Proceedings of the National Academy*, by mid-March at the latest Linus' paper would be spread around the world. Then it would be only a matter of days before the error would be discovered. We had anywhere up to six weeks before Linus again was in full-time pursuit of DNA.

Though Maurice had to be warned, we did not immediately ring him. The pace of Francis' words might cause Maurice to find a reason for terminating the conversation before all the implications of Pauling's folly could be hammered home. Since in several days I was to go up to London to see Bill Hayes, the sensible course was to bring the manuscript with me for Maurice's and Rosy's inspection.

Then, as the stimulation of the last several hours had made further work that day impossible, Francis and I went over to the Eagle. The moment its doors opened for the evening we were there to drink a toast to the Pauling failure. Instead of sherry, I let Francis buy me a whiskey. Though the odds still appeared against us, Linus had not yet won his Nobel.[5]

[5] It was two years before Pauling won his first Nobel Prize in 1954, for Chemistry. He subsequently won the 1962 Nobel Peace Prize.

Chapter 23

Maurice was busy when, just before four, I walked in with the news that the Pauling model was far off base. So I went down the corridor to Rosy's lab, hoping she would be about. Since the door was already ajar, I pushed it open to see her bending over a lighted box upon which lay an X-ray photograph she was measuring. Momentarily startled by my entry, she quickly regained her composure and, looking straight at my face, let her eyes tell me that uninvited guests should have the courtesy to knock.

I started to say that Maurice was busy, but before the insult was out I asked her whether she wanted to look at Peter's copy of his father's manuscript. Though I was curious how long she would take to spot the error, Rosy was not about to play games with me. I immediately explained where Linus had gone astray. In doing so, I could not refrain from pointing out the su-

Rosalind Franklin in the laboratory, the picture taken while she was still in France.

perficial resemblance between Pauling's three-chain helix and the model that
Francis and I had shown her fifteen months earlier. The fact that Pauling's
deductions about symmetry were no more inspired than our awkward efforts
of the year before would, I thought, amuse her. The result was just the opposite. Instead, she became increasingly annoyed with my recurring references to helical structures. Coolly she pointed out that not a shred of evidence
permitted Linus, or anyone else, to postulate a helical structure for DNA.
Most of my words to her were superfluous, for she knew that Pauling was
wrong the moment I mentioned a helix.

Interrupting her harangue, I asserted that the simplest form for any regular polymeric molecule was a helix. Knowing that she might counter with
the fact that the sequence of bases was unlikely to be regular, I went on
with the argument that, since DNA molecules form crystals, the nucleotide
order must not affect the general structure. Rosy by then was hardly able
to control her temper, and her voice rose as she told me that the stupidity
of my remarks would be obvious if I would stop blubbering and look at her
X-ray evidence.

I was more aware of her data than she realized. Several months earlier
Maurice had told me the nature of her so-called antihelical results. Since
Francis had assured me that they were a red herring, I decided to risk a full
explosion. Without further hesitation I implied that she was incompetent in
interpreting X-ray pictures. If only she would learn some theory, she would
understand how her supposed antihelical features arose from the minor dis-
tortions needed to pack regular helices into a crystalline lattice.[1]

Suddenly Rosy came from behind the lab bench that separated us and
began moving toward me. Fearing that in her hot anger she might strike me,
I grabbed up the Pauling manuscript and hastily retreated to the open door.
My escape was blocked by Maurice, who, searching for me, had just then
stuck his head through. While Maurice and Rosy looked at each other over
my slouching figure, I lamely told Maurice that the conversation between
Rosy and me was over and that I had been about to look for him in the tea
room. Simultaneously I was inching my body from between them, leaving
Maurice face to face with Rosy. Then, when Maurice failed to disengage

¹ *Franklin and helical DNA*

Franklin's "antihelical" views had most memorably been demonstrated the previous July when she had penned the tongue-in-cheek notice announcing the "Death of the DNA helix." The black-bordered funeral card, signed by both Franklin and Gosling, was given to Wilkins and Stokes, who did not find it amusing.

There has been much debate about the extent to which Franklin believed DNA not to be helical. The card specifically referred to the A-form (crystalline) DNA, whose structure she largely focused on. Only in February 1953 did she look again at the B-form data she and Gosling had obtained many months earlier and set aside. From her notes at that time it is clear she believed B-form DNA was in fact helical (see Chapter 28). But in the seminar she gave at King's at the end of January 1953, she still said nothing about B DNA, helices, or her data from photo 51. Her focus on the A form, and the apparently non-helical evidence it provided, distracted her from considering helices through 1952.

Crick had also told Franklin she was misinterpreting the asymmetries in her

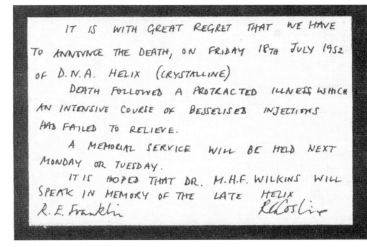

Postcard sent by Franklin and Gosling, announcing the death of DNA helix, July 18, 1952.

data of the A form to suggest it was not a helix. This brief conversation occurred in a queue at a meeting in the Zoological Museum in Cambridge a few months earlier. As reported by Horace Judson and Brenda Maddox, Crick later said: "I'm afraid we always used to adopt—let's say a *patronizing* attitude towards her. When she told us DNA couldn't be a helix, we said, 'Nonsense.' And when she said but her measurements showed that it couldn't, we said, 'Well, they're wrong.'" Despite his confidence, Crick didn't get a good look at the A form photo until after the structure was done. Writing to Wilkins on June 5, 1953, he admitted: "This is the first time I have had an opportunity for a detailed study of the picture of structure A, and I must say I am glad I didn't see it earlier, as it would have worried me considerably."

himself immediately, I feared that out of politeness he would ask Rosy to join us for tea. Rosy, however, removed Maurice from his uncertainty by turning around and firmly shutting the door.

Walking down the passage, I told Maurice how his unexpected appearance might have prevented Rosy from assaulting me. Slowly he assured me that this very well might have happened. Some months earlier she had made a similar lunge toward him. They had almost come to blows following an argument in his

[3] On the right, a page from a notebook recording DNA diffraction experiments carried out at King's in 1952, including some performed by Rosalind Franklin on the transition of Signer DNA between the A (crystalline) and B (wet) forms. Plate 578 taken by Maurice Wilkins was an X-ray diffraction photograph of intact *Sepia* sperm heads, looking at the configuration of DNA and chromosomes in a more natural state. Note that Franklin is referred to as *"Miss* Franklin" while Wilkins is *"Dr.* Wilkins" as was common practice at the time.

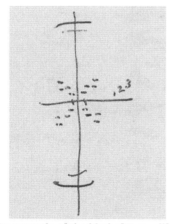

[4] An example of the kind of X-ray diffraction picture Wilkins was taking of *Sepia* sperm is shown in this diagram in a letter to Crick in early 1952. In the same letter, Wilkins refers to the moratorium but writes that he looks "...forward to discussing all our latest ideas & results with you again. Why don't you come & have lunch with me when you are next in town?"

Record of diffraction experiments.

room. When he wanted to escape, Rosy had blocked the door and had moved out of the way only at the last moment. But then no third person was on hand.

My encounter with Rosy opened up Maurice to a degree that I had not seen before. Now that I need no longer merely imagine the emotional hell he had faced during the past two years, he could treat me almost as a fellow collaborator rather than as a distant acquaintance with whom close confidences inevitably led to painful misunderstandings. To my surprise, he revealed that with the help of his assistant Wilson[2] he had quietly been duplicating some of Rosy's and Gosling's X-ray work.[3,4] Thus there need not be a large time gap before Maurice's research efforts were in full swing. Then the even more important cat was let out of the bag: since the middle of the summer Rosy had had evidence for a new three-dimensional form of DNA. It occurred when the DNA molecules were surrounded by a large amount of water. When I asked what the pattern was like, Maurice went into the adjacent room to pick up a print of the new form they called the "B" structure.

The instant I saw the picture my mouth fell open and my pulse began to race. The pattern was unbelievably simpler than those obtained previously ("A" form). Moreover, the black cross of reflections which dominated the picture could arise only from a helical structure. With the A form, the argument for a helix was never straightforward, and considerable ambiguity existed as to exactly which type of helical symmetry was present. With the B form, however, mere inspection of its X-ray picture gave several of the vital helical parameters. Conceivably, after only a few minutes' calculations, the number of chains in the molecule could be fixed.

Herbert Wilson.

Pressing Maurice for what they had done using the B photo, I learned that his colleague R. D. B. Fraser earlier had been doing some serious playing with three-chain models but that so far nothing exciting had come up.[5] Though Maurice conceded that the evidence for a helix was now overwhelming—the Stokes-Cochran-Crick theory clearly indicated that a helix must exist—this was not to him of major significance. After all, he had previously thought a helix would emerge. The real problem was the absence of any structural hypothesis which would allow them to pack the bases regularly in the inside of the helix. Of course this presumed

Bruce Fraser, 1951.

that Rosy had hit it right in wanting the bases in the center and the backbone outside. Though Maurice told me he was now quite convinced she was correct, I remained skeptical, for her evidence was still out of the reach of Francis and me.[6]

On our way to Soho for supper I returned to the problem of Linus, emphasizing that smiling too long over his mistake might be fatal. The position would be far safer if Pauling had been merely wrong instead of looking like a fool. Soon, if not already, he would be at it day and night. There was the

[2] Wilson joined Wilkins at King's in September 1952 on a fellowship from the University of Wales to work on X-ray diffraction of DNA and nucleoproteins. He was a coauthor with Wilkins and Stokes of the second of the papers announcing the structure of DNA in *Nature* in April 1953. Over the next few years he worked with Wilkins on refining the DNA structure before moving to Scotland and working on detailed structures of various components of nucleic acids.

[5] R. D. B. "Bruce" Fraser's model of DNA was never published, despite being referenced in the Watson and Crick paper in April 1953 as "In the Press," a circumstance we shall return to in a later chapter. Fraser's model was a three chain affair, but unlike those of Watson and Crick in 1951 and Pauling in 1953, the phosphate backbone was on the outside and the bases on the inside, holding the chains together through base stacking (not base pairing) interactions.

[6] *The Famous Photograph 51*

Ray Gosling's (2012) account of how Franklin gave photograph 51 to Wilkins:

"Even before the stunning revelation of Watson and Crick's model of NaDNA at the Cavendish, the atmosphere at King's had become one of suppressed turbulence due to the tension between Rosalind and Maurice. Randall had reluctantly taken the view that Rosalind should leave King's. For her part, Rosalind had already had discussions with Bernal, at Birkbeck, with a view to her working on the structure of tobacco mosaic virus. Randall made possible the smooth transfer of her Turner–Newall fellowship and set March 1953 as the time for her departure.

Rosalind and I were working hard on calculating the Patterson function of the A structure, despite Randall's edict that she should cease all work on DNA. This was an impossible ban since we had so much to write up. Indeed, we were putting the finishing touches to two papers to *Acta Crystallographica* in January 1953. It was then that Rosalind realised, with regard to structure B, that she would not have time to go beyond the draft analysis which we had already started and was published as the third paper in the triumvirate in *Nature* April 1953. She therefore decided to make a "present" to Maurice of the original film of our best structure B diffraction pattern, the 51st exposure in our series of X-rays of single

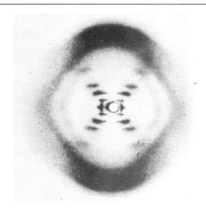

Photograph 51, the B-form of DNA. This is the picture Wilkins showed Watson, page 181.

fibre specimens held at various steady humidities.

Accordingly, I went down the corridor to Maurice's lab/office, sometime in January 1953 and gave him this beautiful negative. He was very surprised and wanted reassurance that Rosalind was actually saying he could make whatever use he wished of this interesting data. This, of course, confirmed the belief of Alex Stokes and himself that the structure was helical. In this respect, after sight of the Watson and Crick model, Rosalind changed her mind, especially in view of the X-ray equivalence of the hydrogen bonded specific paired bases, and admitted that structure A must also contain helical units.

In spite of Randall's edict, Rosalind and I set about making a vector difference map with our Patterson data, which to our great satisfaction and delight, fitted well with a two chain helical symmetric unit."

In his autobiography, Wilkins describes how he was given photograph 51.

"One day in January [30] 1953, Raymond met me in the corridor and handed me an excellent B pattern that Rosalind and he had taken. For me to be shown raw data in this way was quite without precedent and, even more extraordinary, Raymond made it clear that I was to keep the photograph…I had recently been relieved to hear that Rosalind was going to leave our lab for a post at Birkbeck College, and was finishing up her work. I assumed that my being shown the pattern was connected with her plans to leave, and she was handing over the data so that we could follow up on what she and Raymond had done… Raymond gave me to understand that Rosalind was handing the pattern over to me to use as I wished."

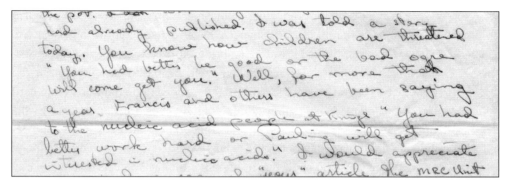

Peter Pauling's letter to Linus, January 13, 1953.

further danger that if he put one of his assistants to taking DNA photographs, the B structure would also be discovered in Pasadena. Then, in a week at most, Linus would have the structure.[7]

Maurice refused to get excited. My repeated refrain that DNA could fall at any moment sounded too suspiciously like Francis in one of his over-wrought periods. For years Francis had been trying to tell him what was important, but the more dispassionately he considered his life, the more he knew he had been wise to follow up his own hunches. As the waiter peered over his shoulder, hoping we would finally order, Maurice made sure I understood that if we could all agree where science was going, everything would be solved and we would have no recourse but to be engineers or doctors.

With the food on the table I tried to fix our thoughts on the chain number, arguing that measuring the location of the innermost reflection on the first and second layer lines might immediately set us on the right track. But since Maurice's long-drawn-out reply never came to the point, I could not decide whether he was saying that no one at King's had measured the pertinent reflections or whether he wanted to eat his meal before it got cold. Reluctantly I ate, hoping that after coffee I might get more details if I walked him back to his flat. Our bottle of Chablis, however, diminished my desire for hard facts, and as we walked out of Soho and across Oxford Street, Maurice spoke only of his plans to get a less gloomy apartment in a quieter area.[8]

[7] Peter Pauling tells his father how Watson and Crick have been trying to get the King's group moving on DNA: "…I was told a story today. You know how children are threatened 'You had better be good or the bad ogre will come get you.' Well, for more than a year Francis and others have been saying to the nucleic acid people at King's 'You had better work hard or Pauling will get interested in nucleic acids…'" (January 13, 1953)

[8] Wilkins finally found a new flat a few months later, reported in a letter to Crick of June 3, 1953. This is presumably the flat in 59 Great Cumberland Place referred to in the MI5 document shown in Chapter 17.

Afterwards, in the cold, almost unheated train compartment, I sketched on the blank edge of my newspaper what I remembered of the B pattern. Then as the train jerked toward Cambridge, I tried to decide between two- and three-chain models. As far as I could tell, the reason the King's group did not like two chains was not foolproof. It depended upon the water content of the DNA samples, a value they admitted might be in great error. Thus by the time I had cycled back to college and climbed over the back gate, I had decided to build two-chain models. Francis would have to agree. Even though he was a physicist, he knew that important biological objects come in pairs.[9]

Chapter 24

Not reading Nature, *here Crick sits in bed with his daughters, Gabrielle to his right and Jacqueline on his left.*

Bragg was in Max's office when I rushed in the next day to blurt out what I had learned. Francis was not yet in, for it was a Saturday morning and he was still home in bed glancing at the *Nature* that had come in the morning mail. Quickly I started to run through the details of the B form, making a rough sketch to show the evidence that DNA was a helix which repeated its pattern every 34 Å along the helical axis. Bragg soon interrupted me with a question, and I knew my argument had got across. I thus wasted no time in bringing up the problem of Linus, giving the opinion that he was far too dangerous to be allowed a second crack at DNA while the people on this side of the Atlantic sat on their hands. After saying that I was going to ask a Cavendish machinist to make models of the purines and pyrimidines, I remained silent, waiting for Bragg's thoughts to congeal.

To my relief, Sir Lawrence not only made no objection but encouraged me to get on with the job of building models. He clearly was not in sympathy with the internal squabbling at King's—especially when it might allow Linus, of all people, to get the thrill of discovering the structure of still another important molecule. Also aiding our cause was my work on tobacco mosaic virus. It had given Bragg the impression that I was on my own. Thus he could fall asleep that night untroubled by the nightmare that he had given Crick

The Cavendish machine shop.

carte blanche for another foray into frenzied inconsiderateness. I then dashed down the stairs to the machine shop to warn them that I was about to draw up plans for models wanted within a week.

[1] Watson had written to his sister concerning her stay in Cambridge on December 11, 1952. "I have just come back from a French lesson with Camille Prior...I talked with her about finding you a place to stay in Cambridge. She has very kindly offered to put you up at her home or if this is not possible for all of your stay, at least to arrange for you to eat in her home." As it turned out, although she did take her meals at Prior's, at least part of the time Elizabeth lodged with Frances Cornford on Millington Road. A granddaughter of Charles Darwin, Cornford was a well-known poet (her most famous poem was *To a Fat Lady Seen from a Train*) and her first cousin, Gwen Raverat, would soon publish a memoir, *Period Piece: A Cambridge Childhood*, dedicated to Frances.

Shortly after I was back in our office, Francis strolled in to report that their last night's dinner party was a smashing success. Odile was positively enchanted with the French boy that my sister had brought along. A month previously Elizabeth had arrived for an indefinite stay on her way back to the States. Luckily I could both install her in Camille Prior's boarding house and arrange to take my evening meals there with Pop and her foreign girls. Thus in one blow Elizabeth had been saved from typical English digs, while I looked forward to a lessening of my stomach pains.[1]

Also living at Pop's was Bertrand Fourcade, the most beautiful male, if not person, in Cambridge. Bertrand, then visiting for a few months to perfect his English, was not unconscious of his unusual beauty and so welcomed the companionship of a girl whose dress was not in shocking contrast with his well-cut clothes. As soon as I had mentioned that we knew the handsome foreigner, Odile expressed delight. She, like many Cambridge women, could not take her eyes off Bertrand whenever she spotted him walking down King's Parade or standing about looking very well-favored during the intermissions of plays at the amateur dramatic club. Elizabeth was thus given the task of seeing whether Bertrand would be free to join us for a meal with the Cricks at Portugal Place. The time finally arranged, however, had overlapped my visit to London. When I was watching Maurice meticulously finish all the food on his plate, Odile was admiring Bertrand's perfectly proportioned face

Odile Crick, Watson, and Elizabeth, Cambridge, 1953.

Bertrand Fourcade with friends.

[2] The "most beautiful male" in Cambridge, Bertrand Fourcade is pictured here, enjoying his summer with Countess Christina Paolozzi and the Australian model Maggi Eckardt, in a photograph taken by Pierre Boulat in 1963, 10 years after the events Watson describes. Bertrand was in Cambridge to get a certificate of English literacy enabling him to enter Harvard Business School. After a period at *Newsweek*, he joined *Vogue* in Paris as advertising director. Bertrand had three brothers. Vincent was a legendary interior designer living in New York and Paris with his partner Robert Denning. Their business embraced the flamboyant excesses of 1980s style— "Outrageous luxury is what our clients want. We have taught them to prefer excess." Xavier was a contemporary art dealer based in New York who knew many of the artists he represented, including William de Kooning. A third brother, Dominique, is a poet and art critic.

as he spoke of his problems choosing among potential social engagements during his forthcoming summer on the Riviera.[2]

This morning Francis saw that I did not have my usual interest in the French moneyed gentry. Instead, for a moment he feared that I was going to be unusually tiresome. Reporting that even a former birdwatcher could now solve DNA was not the way to greet a friend bearing a slight hangover. However, as soon as I revealed the B-pattern details, he knew I was not pulling his leg. Especially important was my insistence that the meridional reflection at 3.4 Å was much stronger than any other reflection. This could only mean that the 3.4 Å-thick purine and pyrimidine bases were stacked on top of each other in a direction perpendicular to the helical axis. In addition we could feel sure from both electron-microscope and X-ray evidence that the helix diameter was about 20 Å.

Francis, however, drew the line against accepting my assertion that the repeated finding of twoness in biological systems told us to build two-chain models. The way to get on, in his opinion, was to reject any argument which did not arise from the chemistry of nucleic-acid chains. Since the experimental evidence known to us could not yet distinguish between two and three-chain models, he wanted to pay equal attention to both alternatives. Though I remained totally skeptical, I saw no reason to contest his words. I would of course start playing with two-chain models.

No serious models were built, however, for several days. Not only did we lack the purine and pyrimidine components, but we had never had the shop put together any phosphorus atoms. Since our machinist needed at least three days merely to turn out the more simple phosphorus atoms, I went back to Clare after lunch to hammer out the final draft of my genetics manuscript. Later, when I cycled over to Pop's for dinner, I found Bertrand and my sister talking to Peter Pauling, who the week before had charmed Pop into giving him dining rights. In contrast to Peter, who was complaining that the Perutzes had no right to keep Nina home on a Saturday night, Bertrand and Elizabeth looked pleased with themselves. They had just returned from motoring in a friend's Rolls to a celebrated country house near Bedford.[3] Their host, an antiquarian architect, had never truckled under to modern civilization and kept his house free of gas and electricity. In all ways possible he maintained the life of an eighteenth-century squire, even to providing special walking sticks for his guests as they accompanied him around his grounds.[4]

Dinner was hardly over before Bertrand whisked Elizabeth on to another party, leaving Peter and me at a loss for something to do. After first deciding to work on his hi-fi set, Peter came along with me to a film. This kept us in check until, as midnight approached, Peter held forth on how Lord Rothschild

[3] The Rolls-Royce belonged to Sri Lankan Geoffrey Bawa who spent one year at Cambridge before moving to London to enroll as a student at the Architectural Association. After returning to Sri Lanka, Bawa became a world-famous architect, founder of "tropical modernism."

[4] "Their host" was Albert Edward Richardson, an architect whose primary love was the architecture of the late Georgian period. He was a President of the Royal Academy and appointed a Knight Commander of the Victorian Order in 1956. His home was Avenue House in the village of Ampthill, Bedfordshire. As Watson writes, it did not have electricity so that he could experience, in part at least, what life was like in Georgian times. Not surprisingly, he protested (in vain) against the installation of modern street lighting in Ampthill.

Richardson with his sign protesting street lights in his village.

was avoiding his responsibility as a father by not inviting him to dinner with his daughter Sarah. I could not disagree, for if Peter moved into the fashionable world I might have a chance to escape acquiring a faculty-type wife.[5]

Three days later the phosphorus atoms were ready, and I quickly strung together several short sections of the sugar-phosphate backbone. Then for a day and a half I tried to find a suitable two-chain model with the backbone in the center. All the possible models compatible with the B-form

Victor Rothschild, 1965.

X-ray data, however, looked stereochemically even more unsatisfactory than our three-chained models of fifteen months before. So, seeing Francis absorbed by his thesis, I took off the afternoon to play tennis with Bertrand. After tea I returned to point out that it was lucky I found tennis more pleasing than model building. Francis, totally indifferent to the perfect spring day, immediately put down his pencil to point out that not only was DNA very important, but he could assure me that someday I would discover the unsatisfactory nature of outdoor games.

During dinner at Portugal Place I was back in a mood to worry about what was wrong. Though I kept insisting that we should keep the backbone in the center, I knew none of my reasons held water. Finally over coffee I admitted that my reluctance to place the bases inside partially arose from the suspicion that it would be possible to build an almost infinite number of models of this type. Then we would have the impossible task of deciding whether one was right. But the real stumbling block was the bases. As long as they were outside, we did not have to consider them. If they were pushed inside, the frightful problem existed of how to pack together two or more chains with irregular sequences of bases. Here Francis had to admit that he saw not the slightest ray of light. So when I walked up out of their basement dining room into the street, I left Francis with the impression that he would have to provide at least a semiplausible argument before I would seriously play about with base-centered models.

The next morning, however, as I took apart a particularly repulsive back-

[5] Victor Rothschild was Nathaniel Mayer Victor Rothschild, 3rd Baron Rothschild. A graduate of Trinity College, Cambridge, he carried out research on fertilization and the physiology of spermatozoa, working with Murdoch Mitchison (pictured on page 107). During the Second World War, Rothschild worked in military intelligence and was awarded the George Medal. He played cricket for Northamptonshire. As an undergraduate at Trinity, he was close friends with Kim Philby, Donald McLean, Guy Burgess, and Anthony Blunt, all found to have been Soviet spies. It was known that there was a fifth spy and Rothschild himself came under suspicion in the 1970s. In 1980, the "Fifth Man" was revealed publicly to be John Cairncross, another undergraduate at Trinity.

bone-centered molecule, I decided that no harm could come from spending a few days building backbone-out models. This meant temporarily ignoring the bases, but in any case this had to happen since now another week was required before the shop could hand over the flat tin plates cut in the shapes of purines and pyrimidines.

There was no difficulty in twisting an externally situated backbone into a shape compatible with the X-ray evidence. In fact, both Francis and I had the impression that the most satisfactory angle of rotation between two adjacent bases was between 30 and 40 degrees. In contrast, an angle either

A sketch of Maurice Wilkins by Hugo "Puck" Dachinger, 1980.

twice as large or twice as small looked incompatible with the relevant bond angles. So if the backbone was on the outside, the crystallographic repeat of 34 Å had to represent the distance along the helical axis required for a complete rotation. At this stage Francis' interest began to perk up, and at increasing frequencies he would look up from his calculations to glance at the model. Nonetheless, neither of us had any hesitation in breaking off work for the weekend. There was a party at Trinity on Saturday night, and on Sunday Maurice was coming up to the Cricks' for a social visit arranged weeks before the arrival of the Pauling manuscript.

Maurice, however, was not allowed to forget DNA. Almost as soon as he arrived from the station, Francis started to probe him for fuller details of the B pattern. But by the end of lunch Francis knew no more than I had picked up the week before. Even the presence of Peter, saying he felt sure his father would soon spring into action, failed to ruffle Maurice's plans. Again he emphasized that he wanted to put off more model building until Rosy was gone, six weeks from then. Francis seized the occasion to ask Maurice whether he would mind if we started to play about with DNA models. When Maurice's slow answer emerged as no, he wouldn't mind, my pulse rate returned to normal. For even if the answer had been yes, our model building would have gone ahead.[6]

[6] Wilkins reflected on this key episode in his autobiography. When Watson and Crick asked him if they could work on DNA again, he "...found their question horrible...But when I assessed the extent of the log-jam in our DNA work at King's, it seemed obvious that I could not ask Francis and Jim to hold off model-building any longer...DNA was not private property: it was open to all to study peacefully without any one person throwing his weight about. I could see no alternative but to accept their position—I had principles and science had to march on. But I was very cast down and could not conceal it."

Chapter 25

The next few days saw Francis becoming increasingly agitated by my failure to stick close to the molecular models. It did not matter that before his tenish entrance I was usually in the lab. Almost every afternoon, knowing that I was on the tennis court, he would fretfully twist his head away from his work to see the polynucleotide backbone unattended.[1] Moreover, after tea I would show up for only a few minutes of minor fiddling before dashing away to have sherry with the girls at Pop's. Francis' grumbles did not disturb me, however, because further refining of our latest backbone without a solution to the bases would not represent a real step forward.

I went ahead spending most evenings at the films, vaguely dreaming that any moment the answer would suddenly hit me. Occasionally my wild pursuit of the celluloid backfired, the worst occasion being an evening set aside for *Ecstasy*. Peter and I had both been too young to observe the original showings of Hedy Lamarr's romps in the nude, and so on the long-awaited night we collected Elizabeth and went up to the Rex.[2] However, the only swimming scene left intact by the English censor was an inverted reflection from a pool of water.

[1] Watson's enthusiasm for tennis was indeed keeping him out of the lab on many afternoons. In letters to his sister his time on the court is often mentioned. On April 27 the previous year, for example: "I have started to play tennis rather regularly—3 times in the last week." On July 8: "I have played tennis several times in the last few days, and to my surprise rather well...It is very satisfying to see a well driven shot from my backhand."

[2] The Rex cinema was a "flea pit" popular among Cambridge students for the mix of art house and classic films it showed under the management of the recently graduated Leslie Halliwell who would later become well known for his published Film Guides.

The Rex cinema.

[3] This 1933 Czech film starred the young Hedy Lamarr (then still known as Zvonimir Rogoz). Disappointment at the deleted swimming scenes was understandable: it was largely Lamarr's nude swimming and running naked through a wood that had earned the movie its controversial fame.

Before the film was half over we joined the violent booing of the disgusted undergraduates as the dubbed voices uttered words of uncontrolled passion.[3]

Even during good films I found it almost impossible to forget the bases. The fact that we had at last produced a stereochemically reasonable configuration for the backbone was always in the back of my head. Moreover, there was no longer any fear that it would be incompatible with the experimental data. By then it had been checked out with Rosy's precise measurements. Rosy, of course, did not directly give us her data. For that matter, no one at King's realized they were in our hands. We came upon them because of Max's membership on a committee appointed by the Medical Research Council to look into the research activities of Randall's lab. Since Randall wished to convince the outside committee that he had a productive research group, he had instructed his people to draw up a comprehensive summary of their accomplishments. In due time this was prepared in mimeograph form and sent routinely to all the committee members. As soon as Max saw the sections by Rosy and Maurice, he brought the report in to Francis and me. Quickly scanning its contents, Francis sensed with relief that following my return from King's I had correctly reported to him the essential features of the B pattern. Thus only minor modifications were necessary in our backbone configuration.[4,5]

Generally, it was late in the evening after I got back to my rooms that I tried to puzzle out the mystery of the bases. Their formulas were written out in J. N. Davidson's little book *The Biochemistry of Nucleic Acids*, a copy of which I kept in Clare. So I could be sure that I had the correct structures when I drew tiny pictures of the bases on sheets of Cavendish notepaper. My aim was somehow to arrange the centrally located bases in such a way that the backbones on the outside were completely regu-

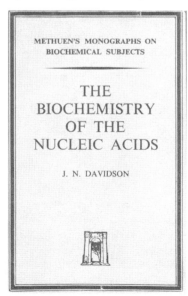

Davidson's classic textbook on nucleic acids.

MEDICAL RESEARCH COUNCIL

MRC.53/74

BIOPHYSICS COMMITTEE

The twelfth meeting of the Biophysics Committee was held at the Biophysics Research Unit, King's College, London, on Monday, the 15th December, 1952, at 12 noon.

Present:

Sir Edward Salisbury (Chairman)
Professor W.T. Astbury
Dr. Honor Fell
Dr. L.H. Gray
Dr. E.R. Holiday
Dr. A.S. McFarlane
Dr. M.F. Perutz
Professor J.T. Randall
Mr. J.W. Boag   (Secretary)

Apologies for absence were received from:-

Dr. R.B. Bourdillon
Sir Charles Lovatt Evans
Sir Charles Harington
Professor J.S. Mitchell
Sir Rudolph Peters
Dr. E.E. Pochin

Sir John

It seems to me that it is more suitable that you, rather than I, reply to Perutz concerning the question of how proper it was for him to pass on the Report to Watson and Crick.   My guess is that you would have sent a rather different Report to Head Office if you had known that the members of the Committee were going to hand it round generally?   If Perutz thinks that the only documents one should not show to other people are those marked "Restricted" or "Confidential", he seems to me to be living in a funny world!

I enclose a draft copy of my reply to Perutz.

I wonder how many more repercussions there will be of the Watson book?

MHFW
19 December '68

---

[4] The front page of the report prepared by Randall for the MRC Biophysics Committee which reviewed the work of his MRC unit on December 15, 1952. Members of his unit contributed descriptions of their work.

The most significant new fact gleaned by Watson and Crick from the MRC report was the space group (see box on page 196). Max Perutz showing the MRC report to Watson and Crick was severely criticized when *The Double Helix* was published. In particular, Chargaff, in his review of the book published in *Science*, claimed that Perutz had improperly shown Watson and Crick a "confidential" report. This review triggered letters to *Science* from Perutz, Wilkins, and Watson clarifying their views of the incident. Therein they claimed that the MRC report had been compiled by Randall for a committee set up to ensure sharing of information between MRC units, and that it was not marked, nor understood to be, confidential. Perutz does however concede that, as a matter of courtesy, he should have asked Randall's permission to show the particulars to Watson and Crick. The Chargaff review and responding letters are reproduced in Appendix 5.

Wilkins' opinion of Perutz' action in giving the MRC report to Watson and Crick is clear in this memorandum to John Randall, December 19, 1968.

lar—that is, giving the sugar phosphate groups of each nucleotide identical three-dimensional configurations. But each time I tried to come up with a solution I ran into the obstacle that the four bases each had a quite different shape. Moreover, there were many reasons to believe that the sequences of the bases of a given polynucleotide chain were very irregular. Thus, unless some very special trick existed, randomly twisting two polynucleotide chains around one another should result in a mess. In some places the bigger bases must touch each other, while in other regions, where the smaller bases would lie opposite each other, there must exist a gap or else their backbone regions must buckle in.

There was also the vexing problem of how the intertwined chains might be held together by hydrogen bonds between the bases. Though for over a year Francis and I had dismissed the possibility that bases formed regular hydrogen bonds,

James Michael Creeth.

D. O. Jordan.

it was now obvious to me that we had done so incorrectly. The observation that one or more hydrogen atoms on each of the bases could move from one location to another (a tautomeric shift) had initially led us to conclude that all the possible tautomeric forms of a given base occurred in equal frequencies. But a recent rereading of J. M. Gulland's and D. O. Jordan's papers on the acid and base titrations of DNA made me finally appreciate the strength of their conclusion that a large fraction, if not all, of the bases formed hydrogen bonds to other bases.[6]

[6] This work, begun by J. M. Gulland and D. O. Jordan at University College Nottingham, culminated with an experiment performed by their young graduate student, Mike Creeth. The paper was published in 1947.

The work of the Nottingham group sadly ended there, as Gulland was killed along with 27 other people when the 11:15 a.m. express train from Edinburgh to London on October 26, 1947, derailed at about 12:45 p.m. at Goswick, 6 miles south of Berwick.

Train crash in which J. M. Gulland died.

J. M. Gulland.

Even more important, these hydrogen bonds were present at very low DNA concentrations, strongly hinting that the bonds linked together bases in the same molecule. There was in addition the X-ray crystallographic result that each pure base so far examined formed as many irregular hydrogen bonds as stereochemically possible. Thus, conceivably the crux of the matter was a rule governing hydrogen bonding between bases.

My doodling of the bases on paper at first got nowhere, regardless of whether or not I had been to a film. Even the necessity to expunge *Ecstasy* from my mind did not lead to passable hydrogen bonds, and I fell asleep hoping that an undergraduate party the next afternoon at Downing would be full of pretty girls. But my expectations were dashed as soon as I arrived to spot a group of healthy hockey players and several pallid debutantes. Bertrand also instantly perceived he was out of place, and as we passed a polite interval before scooting out, I explained how I was racing Peter's father for the Nobel Prize.

Not until the middle of the next week, however, did a nontrivial idea emerge. It came while I was drawing the fused rings of adenine on paper. Suddenly I realized the potentially profound implications of a DNA structure in which the adenine residue formed hydrogen bonds similar to those found in crystals of pure adenine. If DNA was like this, each adenine residue would form two hydrogen bonds to an adenine residue related to it by a 180-degree rotation. Most important, two symmetrical hydrogen bonds could also hold together pairs of guanine, cytosine, or thymine. I thus started wondering whether each DNA molecule consisted of two chains with identical base sequences held together by hydrogen bonds between pairs of identical bases. There was the complication, however, that such a structure could not have a regular backbone, since the purines (adenine and guanine) and the pyrimidines (thymine and cytosine) have different shapes. The resulting backbone would have to show minor in-and-out buckles depending upon whether pairs of purines or pyrimidines were in the center.

Despite the messy backbone, my pulse began to race. If this was DNA, I should create a bombshell by announcing its discovery. The existence of two intertwined chains with identical base sequences could not be a chance matter. Instead it would strongly suggest that one chain in each molecule

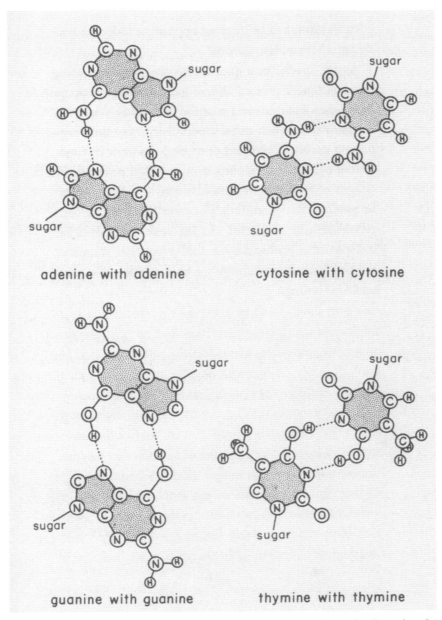

The four base pairs used to construct the like-with-like structure (hydrogen bonds are dotted).

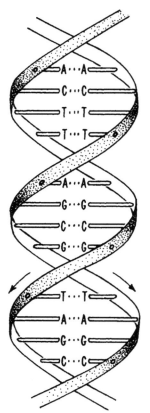

A schematic view of a DNA molecule built up from like-with-like base pairs.

had at some earlier stage served as the template for the synthesis of the other chain. Under this scheme, gene replication starts with the separation of its two identical chains. Then two new daughter strands are made on the two parental templates, thereby forming two DNA molecules identical to the original molecule. Thus, the essential trick of gene replication could come from the requirement that each base in the newly synthesized chain always hydrogen-bonds to an identical base. That night, however, I could not see why the common tautomeric form of guanine would not hydrogen-bond to adenine. Likewise, several other pairing mistakes should also occur. But since there was no reason to rule out the participation of specific enzymes, I saw no need to be unduly disturbed. For example, there might exist an enzyme specific for adenine that caused adenine always to be inserted opposite an adenine residue on the template strands.

As the clock went past midnight I was becoming more and more pleased. There had been far too many days when Francis and I worried that the DNA structure might turn out to be superficially very dull, suggesting nothing about either its replication or its function in controlling cell biochemistry. But now, to my delight and amazement, the answer was turning out to be profoundly interesting. For over two hours I happily lay awake with pairs of adenine residues whirling in front of my closed eyes. Only for brief moments did the fear shoot through me that an idea this good could be wrong.

Chapter 26

My scheme was torn to shreds by the following noon. Against me was the awkward chemical fact that I had chosen the wrong tautomeric forms of guanine and thymine. Before the disturbing truth came out, I had eaten a hurried breakfast at the Whim, then momentarily gone back to Clare to reply to a letter from Max Delbrück which reported that my manuscript on bacterial genetics looked unsound to the Cal Tech geneticists. Nevertheless, he would accede to my request that he send it to the *Proceedings of the National Academy*. In this way, I would still be young when I committed the folly of publishing a silly idea. Then I could sober up before my career was permanently fixed on a reckless course.[1]

At first this message had its desired unsettling effect. But now, with my spirits soaring on the possibility that I had the self-duplicating structure, I re-iterated my faith that I knew what happened when bacteria mated. More-over, I could not refrain from adding a sentence saying that I had just devised a beautiful DNA structure which was completely different from Pauling's. For a few seconds I considered giving some details of what I was up to, but since I was in a rush I decided not to, quickly dropped the letter in the box, and dashed off to the lab.

[1] Despite having reservations, Delbrück communicated the manuscript to E. B. Wilson, editor of *Proceedings of the National Academy of Sciences*, on February 25, 1953.

Delbrück sent Watson a copy of the submission letter with a postscript shown here telling Watson that Alfred Sturtevant and Marguerite Vogt had doubts about the paper. Delbrück concluded:

"However, since you don't want to change it, and since I want to do experiments rather than to rewrite your paper, and since it will do you good to learn what it means to publish prematurely, I sent it off today with only a few commas and missing words amended."

I had started on DNA when I first arrived in Cambridge but had stopped because the Kings group did not like competition or cooperation. However since Pauling is now working on it, I believe the field is open to any body. I thus intend to work on it until the solution is out. Today I am very optimistic since I believe I have a very pretty model, which is so pretty I am surprised no one has thought of it before. When I have the proper coordinates worked out, I shall send a note to Nature, since it accounts for the X-ray data, and even if wrong, is a marked improvement on the Pauling model. I shall send you a copy of the note.

A paragraph from Watson's letter to Delbrück (February 20, 1953) in which he reports that he has "...a very pretty model" for DNA.

The letter was not in the post for more than an hour before I knew that my claim was nonsense. I no sooner got to the office and began explaining my scheme than the American crystallographer Jerry Donohue protested that the idea would not work. The tautomeric forms I had copied out of David-son's book were, in Jerry's opinion, incorrectly assigned. My immediate re-tort that several other texts also pictured guanine and thymine in the enol form cut no ice with Jerry. Happily he let out that for years organic chemists had been arbitrarily favoring particular tautomeric forms over their alternatives on only the flimsiest of grounds. In fact, organic-chemistry textbooks were littered with pictures of highly improbable tautomeric forms. The guanine picture I was thrusting toward his face was almost certainly bogus. All his chemical intuition told him that it would occur in the keto form. He was just as sure that thymine was also wrongly assigned an enol configuration. Again he strongly favored the keto alternative.[2]

6   THE BIOCHEMISTRY OF THE NUCLEIC ACIDS

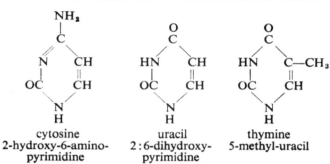

cytosine
2-hydroxy-6-amino-
pyrimidine

uracil
2:6-dihydroxy-
pyrimidine

thymine
5-methyl-uracil

2.3 *Purine bases*

Both types of nucleic acids contain the same purine bases, adenine and guanine. They are derivatives of the parent compound purine which is formed by the fusion of a pyrimidine ring and an iminazole ring.

Purine

Adenine and guanine have the following structures:

adenine
(6-aminopurine)

guanine
(2-amino-6-hydroxypurine)

Davidson's drawings of the bases in the first edition of Biochemistry of Nucleic Acids.

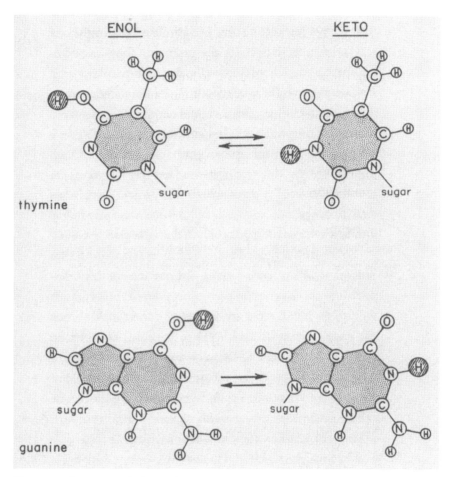

The contrasting tautomeric forms of guanine and thymine which might occur in DNA. The hydrogen atoms that can undergo the changes in position (a tautomeric shift) are shaded.

Jerry, however, did not give a foolproof reason for preferring the keto forms. He admitted that only one crystal structure bore on the problem. This was diketopiperazine, whose three-dimensional configuration had been carefully worked out in Pauling's lab several years before. Here there was no doubt that the keto form, not the enol, was present. Moreover, he felt sure that the quantum-mechanical arguments which showed why diketopiperazine has the keto form should also hold for guanine and thymine. I was thus firmly urged not to waste more time with my harebrained scheme.

Though my immediate reaction was to hope that Jerry was blowing hot air, I did not dismiss his criticism. Next to Linus himself, Jerry knew more about hydrogen bonds than anyone else in the world. Since for many years he had worked at Cal Tech on the crystal structures of small organic molecules, I couldn't kid myself that he did not grasp our problem. During the six months that he occupied a desk in our office, I had never heard him shooting off his mouth on subjects about which he knew nothing.

Thoroughly worried, I went back to my desk hoping that some gimmick might emerge to salvage the like-with-like idea. But it was obvious that the new assignments were its death blow. Shifting the hydrogen atoms to their keto locations made the size differences between the purines and pyrimidines even more important than would be the case if the enol forms existed. Only by the most special pleading could I imagine the polynucleotide backbone bending enough to accommodate irregular base sequences. Even this possibility vanished when Francis came in. He immediately realized that a like-with-like structure would give a 34 Å crystallographic repeat only if each chain had a complete rotation every 68 Å. But this would mean that the rotation angle between successive bases would be only 18 degrees, a value Francis believed was absolutely ruled out by his recent fiddling with the models. Also Francis did not like the fact that the structure gave no explanation for the Chargaff rules (adenine equals thymine, guanine equals cytosine). I, however, maintained my lukewarm response to Chargaff's data. So I welcomed the arrival of lunchtime, when Francis' cheerful prattle temporarily shifted my thoughts to why undergraduates could not satisfy *au pair* girls.

After lunch I was not anxious to return to work, for I was afraid that in trying to fit the keto forms into some new scheme I would run into a stone wall and have to face the fact that no regular hydrogen-bonding scheme was compatible with the X-ray evidence. As long as I remained outside gazing at the crocuses, hope could be maintained that some pretty base arrangement would

Jerry Donohue. His shirt suggests that this photograph was taken in California rather than Cambridge.

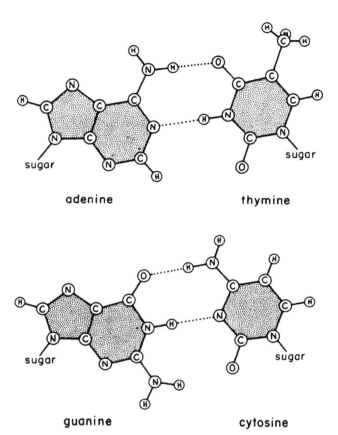

The adenine-thymine and guanine-cytosine base pairs used to construct the double helix (hydrogen bonds are dotted). The formation of a third hydrogen bond between guanine and cytosine was considered, but rejected because a crystallographic study of guanine hinted that it would be very weak. Now this conjecture is known to be wrong. Three strong hydrogen bonds can be drawn between guanine and cytosine.

fall out. Fortunately, when we walked upstairs, I found that I had an excuse to put off the crucial model-building step for at least several more hours. The metal purine and pyrimidine models, needed for systematically checking all the conceivable hydrogen-bonding possibilities, had not been finished on time. At least two more days were needed before they would be in our hands. This was much too long even for me to remain in limbo, so I spent the rest of the afternoon cutting accurate representations of the bases out of stiff cardboard. But by the time they were ready I realized that the answer must be put off till the next day. After dinner I was to join a group from Pop's at the theater.

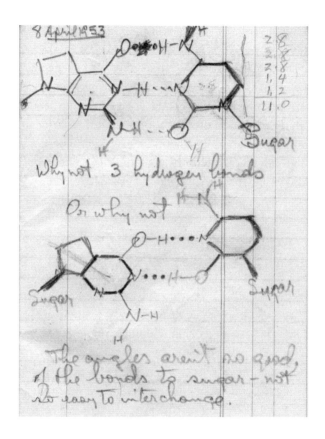

[3] **Watson mentions in the caption to the diagram of the base pairs in** *The Double Helix* **(shown on opposite page) that "…a third hydrogen bond between guanine and cytosine was considered, but rejected…" It was Linus Pauling and Corey who determined that there were indeed three hydrogen bonds between guanine and cytosine. Pauling visited Cambridge in April 1953, and saw the model before going on with Bragg to the Solvay Conference in Stockholm (see Chapter 29). Here is his notebook entry of April 8, 1953 pointing out the likelihood of three bonds.**

When I got to our still empty office the following morning, I quickly cleared away the papers from my desk top so that I would have a large, flat surface on which to form pairs of bases held together by hydrogen bonds. Though I initially went back to my like-with-like prejudices, I saw all too well that they led nowhere. When Jerry came in I looked up, saw that it was not Francis, and began shifting the bases in and out of various other pairing possibilities. Suddenly I became aware that an adenine-thymine pair held together by two hydrogen bonds was identical in shape to a guanine-cytosine pair held together by at least two hydrogen bonds.[3] All the hydrogen bonds seemed to form naturally; no fudging was required to make the two types of base pairs identical in shape. Quickly I called Jerry over to ask him whether this time he had any objection to my new base pairs.

When he said no, my morale skyrocketed, for I suspected that we now had the answer to the riddle of why the number of purine residues exactly equaled the number of pyrimidine residues. Two irregular sequences of bases could be regularly packed in the center of a helix if a purine always hydrogen-bonded to a pyrimidine. Furthermore, the hydrogen-bonding requirement meant that adenine would always pair with thymine, while guanine could pair only with cytosine. Chargaff's rules then suddenly stood out as a consequence of a double-helical structure for DNA. Even more exciting, this type of double helix suggested a replication scheme much more satisfactory than my briefly considered like-with-like pairing. Always pairing adenine with thymine and guanine with cytosine meant that the base sequences of the two intertwined chains were complementary to each other. Given the base sequence of one chain, that of its partner was automatically determined. Conceptually, it was thus very easy to visualize how a single chain could be the template for the synthesis of a chain with the complementary sequence.

Upon his arrival Francis did not get more than halfway through the door before I let loose that the answer to everything was in our hands. Though as a matter of principle he maintained skepticism for a few moments, the similarly shaped A-T and G-C pairs had their expected impact. His quickly pushing the bases together in a number of different ways did not reveal any other way to satisfy Chargaff's rules. A few minutes later he spotted the fact that the two glycosidic bonds (joining base and sugar) of each base pair were systematically related by a diad axis perpendicular to the helical axis. Thus, both pairs could be flipflopped over and still have their glycosidic bonds facing in the same direction. This had the important consequence that a given chain could contain both purines and pyrimidines. At the same time, it strongly suggested that the backbones of the two chains must run in opposite directions.

The question then became whether the A-T and G-C base pairs would easily fit the backbone configuration devised during the previous two weeks. At first glance this looked like a good bet, since I had left free in the center a large vacant area for the bases. However, we both knew that we would not

Commemorative plaque on the Eagle celebrating the discovery of the DNA double helix.

be home until a complete model was built in which all the stereochemically contacts were satisfactory. There was also the obvious fact that the implications of its existence were far too important to risk crying wolf. Thus I felt slightly queasy when at lunch Francis winged into the Eagle to tell everyone within hearing distance that we had found the secret of life.[4]

[4] Crick did not recall "…winging into the Eagle" and declaring that they had found the secret of life, but the line is so good, it hardly matters.

Chapter 27

Francis' preoccupation with DNA quickly became full-time. The first afternoon following the discovery that A-T and G-C base pairs had similar shapes, he went back to his thesis measurements, but his effort was ineffectual. Constantly he would pop up from his chair, worriedly look at the cardboard models, fiddle with other combinations, and then, the period of momentary uncertainty over, look satisfied and tell me how important our work was. I enjoyed Francis' words, even though they lacked the casual sense of understatement known to be the correct way to behave in Cambridge. It seemed almost unbelievable that the DNA structure was solved,

Crick fiddling with their DNA model.

that the answer was incredibly exciting, and that our names would be associated with the double helix as Pauling's was with the alpha helix.

When the Eagle opened at six, I went over with Francis to talk about what must be done in the next few days. Francis wanted no time lost in seeing whether a satisfactory three-dimensional model could be built, since the geneticists and nucleic-acid biochemists should not misuse their time and facilities any longer than necessary. They must be told the answer quickly, so that they could reorient their research upon our work. Though I was equally anxious to build the complete model, I thought more about Linus and the possibility that he might stumble upon the base pairs before we told him the answer.

That night, however, we could not firmly establish the double helix. Until the metal bases were on hand, any model building would be too sloppy to be convincing. I went back to Pop's to tell Elizabeth and Bertrand that Francis and I had probably beaten Pauling to the gate and that the answer would revolutionize biology. Both were genuinely pleased, Elizabeth with sisterly pride, Bertrand with the idea that he could report back to International Society that he had a friend who would win a Nobel Prize. Peter's reaction was equally enthusiastic and gave no indication that he minded the possibility of his father's first real scientific defeat.

Clare Bridge.

The following morning I felt marvelously alive when I awoke. On my way to the Whim I slowly walked toward the Clare Bridge, staring up at the gothic pinnacles of the King's College Chapel that stood out sharply against the spring sky.[1] I briefly stopped and looked over at the perfect Georgian features of the recently cleaned

[1] The oldest surviving bridge over the river Cam, it was built by Thomas Grumbold in 1640.

King's College Chapel and Gibbs Building.

[2] The second oldest building of King's College, after the chapel, which was started in 1446, is the Gibbs Building (*right*), on which work began in 1724. The building was based on a design from the architect James Gibbs after two designs by Nicholas Hawksmoor were rejected by the college.

Gibbs Building, thinking that much of our success was due to the long uneventful periods when we walked among the colleges or unobtrusively read the new books that came into Heffer's Bookstore.[2,3] After contentedly poring over *The Times*, I wandered into the lab to see Francis, unquestionably early, flipping the cardboard base pairs about an imaginary line. As far as a compass and ruler could tell him, both sets of base pairs neatly fitted into the backbone configuration. As the morning wore on, Max and John successively came by to see if we still thought we had it. Each got a quick, concise lecture from Francis, during the second of which I wandered down to see if the shop could be speeded up to produce the purines and pyrimidines later that afternoon.

Only a little encouragement was

THE OAK ROOM AND GALLERY
in our Petty Cury Bookshop

Here, at the back of the building, on the first and second floors, is displayed our extensive stock of Secondhand Books—on English Literature, History, Art, Philosophy, Theology, and so on.   Elsewhere you will find new books of general interest, scientific and technical books, orientalia, &c., &c.

W. HEFFER & SONS LTD
3-4 PETTY CURY, CAMBRIDGE
Telephone 58351

xvii

1953 advertisement for Heffer and Sons bookshop.

[3] By the early 1950s, Heffer's had already been selling books in Cambridge for 75 years. The shop visited by Watson was located in Petty Cury, but moved in the 1970s to its current location in Trinity Street, in the very premises occupied by Matthew & Son Ltd at the time of this story (see Chapter 9).

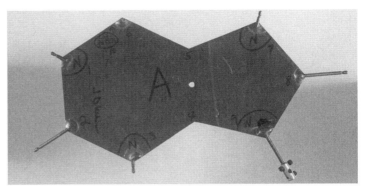

Base plate used in the original model.

4 The original DNA model built by Watson and Crick was later dismantled, but this is one of the metal base plates built in the machine shop at the Cavendish and, it is claimed, used in that model.

5 The DNA helix is right-handed but all too many pictures of DNA display a left-handed helix. It seems that art directors do not know that the left and right helices are topologically different and frequently reverse images of DNA. There is a left-handed DNA, Z-DNA, discovered by Alex Rich (pictured on page 167), but this is a very different structure.

needed to get the final soldering accomplished in the next couple of hours. The brightly shining metal plates were then immediately used to make a model in which for the first time all the DNA components were present.[4] In about an hour I had arranged the atoms in positions which satisfied both the X-ray data and the laws of stereochemistry. The resulting helix was right-handed with the two chains running in opposite directions.[5] Only one person can easily play with a model, and so Francis did not try to check my work until I backed away and said that I thought everything fitted. While one interatomic contact was slightly shorter than optimal, it was not out of line with several published values, and I was not disturbed. Another fifteen minutes' fiddling by Francis failed to find anything wrong, though for brief intervals my stomach felt uneasy when I saw him frowning. In each case he became satisfied and moved on to verify that another interatomic contact was reasonable. Everything thus looked very good when we went back to have supper with Odile.

Our dinner words fixed on how to let the big news out. Maurice, especially, must soon be told. But remembering the fiasco of sixteen months before, keeping King's in the dark made sense until exact coordinates had been obtained for all the atoms. It was all too easy to fudge a successful series of atomic contacts so that, while each looked almost acceptable, the whole collection was energetically impossible. We suspected that we had not made this error, but

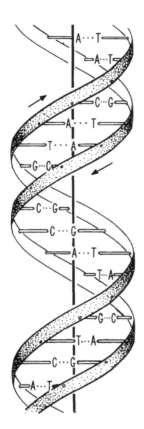

A schematic illustration of the double helix. The two sugar-phosphate backbones twist about on the outside with the flat hydrogen-bonded base pairs forming the core. Seen this way, the structure resembles a spiral staircase with the base pairs forming the steps.

our judgment conceivably might be biased by the biological advantages of complementary DNA molecules. Thus the next several days were to be spent using a plumb line and a measuring stick to obtain the relative positions of all atoms in a single nucleotide. Because of the helical symmetry, the locations of the atoms in one nucleotide would automatically generate the other positions.

After coffee Odile wanted to know whether they would still have to go into exile in Brooklyn if our work was as sensational as everyone told her. Perhaps we should stay on in Cambridge to solve other problems of equal importance. I tried to reassure her, emphasizing that not all American men cut all their hair off and that there were scores of American women who did

6 The Cricks' imminent exile in Brooklyn was because Francis had earlier accepted a position in the lab of David Harker (see Chapter 20). Odile's apprehension was understandable; they were already having trouble finding a suitable apartment, as this letter reveals. And despite Watson's reassurances, the year spent there was not a happy one. Financially hard up, and living in a characterless apartment on the edge of Brooklyn, Odile struggled to look after the family while Crick wrestled with a less conducive work environment than that to which he was accustomed in Cambridge.

```
UNIVERSITY OF CAMBRIDGE    DEPARTMENT OF PHYSICS


                                      CAVENDISH LABORATORY

                                        Free School Lane,
                                             Cambridge.

                                        21st January 1953.

Dr. David Harker,
Polytechnic Institute of Brooklyn,
The Protein Structure Project,
55 Johnson Street, 4th Floor,
Brooklyn 1, New York.


Dear Dr. Harker,

        Thank you for your letter of 7th January.   I have heard
from Wyckoff who seems to think there will be little difficulty
in getting our visa interview rather earlier.    I shall write to
the Embassy as soon as I have cleared Michael with the Divorce Court.
Naturally I did not expect the travelling money before we got the
visa.   I will let you know later on how things can most easily
be arranged.

        As to the apartment, I suggest you relax first the price,
and go higher, say up to $.120.           If there are still
difficulties I think we should not worry too much about the
neighbourhood.   We certainly don't want to buy furniture, and we
really need three bedrooms if this is at all possible.   It is really
very kind of you to do this for us.
```

Letter from Crick to Harker.

not wear short white socks on the streets. I had less success arguing that the States' greatest virtue was its wide-open spaces where people never went. Odile looked in horror at the prospect of being long without fashionably dressed people. Moreover, she could not believe that I was serious, since I had just had a tailor cut a tightly fitting blazer, unconnected with the sacks that Americans draped on their shoulders.[6]

The next morning I again found that Francis had beaten me to the lab. He was already at work tightening the model on its support stands so that he could read off the atomic coordinates. While he moved the atoms back and forth, I sat on the top of my desk thinking about the form of the letters that I soon could write, saying that we had found something interesting. Occasionally, Francis would look disgusted when my daydreams kept me from observing that he needed my help to keep the model from collapsing as he rearranged the supporting ring stands.

By then we knew that all my previous fuss about the importance of Mg^{++} ions was misdirected. Most likely Maurice and Rosy were right in insisting that they were looking at the Na^+ salt of DNA. But with the sugar-phosphate backbone on the outside, it did not matter which salt was present. Either would fit perfectly well into the double helix.

Bragg had his first look late that morning. For several days he had been home with the flu and was in bed when he heard that Crick and I had thought up an ingenious DNA structure which might be important to biology. During his first free moment back in the Cavendish he slipped away from his office for a direct view. Immediately he caught on to the complementary relation between the two chains and saw how an equivalence of adenine with thymine and guanine with cytosine was a logical consequence of the regular repeating shape of the sugar-phosphate backbone. As he was not aware of Chargaff's rules, I went over the experimental evidence on the relative proportions of the various bases, noticing that he was becoming increasingly excited by its potential implications for gene replication. When the question of the X-ray evidence came up, he saw why we had not yet called up the King's group. He was bothered, however, that we had not yet asked Todd's opinion. Telling Bragg that we had got the organic chemistry straight did not put him completely at ease. The chance that we were using the wrong chemical formula admittedly was small, but, since Crick talked so fast, Bragg could never be sure that he would ever slow down long enough to get the right facts. So it was arranged that as soon as we had a set of atomic coordinates, we would have Todd come over.

The final refinements of the coordinates were finished the following evening. Lacking the exact X-ray evidence, we were not confident that the configuration chosen was precisely correct. But this did not bother us, for we only wished to establish that at least one specific two-chain complementary helix was stereochemically possible. Until this was clear, the objection could

[7]On March 7, 1953, Wilkins wrote to Crick:

"I think that you will be interested to know that our dark lady leaves us next week & much of the 3-dimensional data is already in our hands. I am now reasonably clear of other commitments & have started up a general offensive on Nature's secret strongholds on all fronts: models, theoretical chemistry & interpretation of data crystalline & comparative. At last the decks are clear & we can put all hands to the pumps!

It won't be long now.

Regards to all

Yours ever M

P.S. may be in Cambridge next week."

Wilkins alluding to Franklin as "our dark lady" refers to "The Dark Lady" sonnets (127–152) of Shakespeare. The identity of Shakespeare's Dark Lady has been the subject of intense speculation for centuries.

BIOPHYSICS RESEARCH UNIT,
KING'S COLLEGE,
STRAND,
LONDON, W.C.2.
TELEPHONE : TEMPLE BAR 5651

Dark Lady letter.

be raised that, although our idea was aesthetically elegant, the shape of the sugar-phosphate backbone might not permit its existence. Happily, now we knew that this was not true, and so we had lunch, telling each other that a structure this pretty just had to exist.

With the tension now off, I went to play tennis with Bertrand, telling Francis that later in the afternoon I would write Luria and Delbrück about the double helix. It was also arranged that John Kendrew would call up Maurice to say that he should come out to see what Francis and I had just devised. Neither Francis nor I wanted the task. Earlier in the day the post had brought a note from Maurice to Francis, mentioning that he was now about to go full steam ahead on DNA and intended to place emphasis on model building.[7,8]

[8] Crick's response to Wilkins' letter was recorded by Judson (September 10, 1975):

"A remarkable thing," Crick said. "I went to my desk, opened the letter from Maurice, you know, dark lady and this sort of thing, and I looked across and I thought, was it more a question of laughing or—well, you know, sadness almost. You see. There was the model."

Chapter 28

Maurice needed but a minute's look at the model to like it. He had been forewarned by John that it was a two-chain affair, held together by the A-T and G-C base pairs, and so immediately upon entering our office he studied its detailed features. That it had two, not three, chains did not bother him since he knew the evidence never seemed clear-cut. While Maurice silently stared at the metal object, Francis stood by, sometimes talking very fast about what sort of X-ray diagram the structure should produce, then becoming strangely noiseless when he perceived that Maurice's wish was to look at the double helix, not to receive a lecture in crystallographic theory which he could work out by himself. There was no questioning of the decision to put guanine and thymine in the keto form. Doing otherwise would destroy the base pairs, and he accepted Jerry Donohue's spoken argument as if it were a commonplace.[1]

The unforeseen dividend of having Jerry share an office with Francis, Peter, and me, though obvious to all, was not spoken about. If he had not been with us in Cambridge, I might still have been plumping for a like-with-like structure.[2] Maurice, in a lab devoid of structural chemists, did not have anyone about to tell him that all the textbook pictures were wrong. But for Jerry, only Pauling

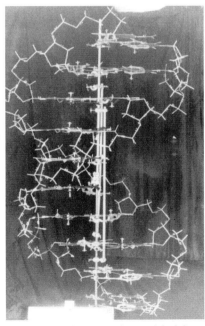

The original demonstration model of the double helix (scale gives distances in Angstroms).

[1] Wilkins described his reaction on seeing the model in his memoir: "But a feeling came through to me that the model, though only bits of wire, on a lab bench, had a special life of its own. It seemed like an incredible new-born baby that spoke for itself, saying 'I don't care what you think—I know I am right.'"

[2] Jerry Donohue appears to have had mixed feelings about his role in the discovery. As late as 1970, he published a paper saying, in effect, that the G-C and A-T base pairing had not been proven. This drew a sharp riposte from Crick: "If Donohue thinks that an equally effective model for DNA could be produced with some alternative base pairing, let him build such a model...it would involve him in a fair amount of work, but I see no other way of deciding the matter."

[3] Wilkins remembered the scene differently. On Watson's insistence, he and Crick had offered Wilkins joint authorship on the paper, an offer that Wilkins refused: "I sounded bitter and Francis said I was unfair. I did not think of thanking Francis and Jim for the generosity of their offer...I firmly believed that what really mattered was scientific progress. I despised thoughts about fame, and perhaps that made me feel insulted by Francis and Jim's generous offer."

would have been likely to make the right choice and stick by its consequences.

The next scientific step was to compare seriously the experimental X-ray data with the diffraction pattern predicted by our model. Maurice went back to London, saying that he would soon measure the critical reflections. There was not a hint of bitterness in his voice, and I felt quite relieved. Until the visit I had remained apprehensive that he would look gloomy, being unhappy that we had seized part of the glory that should have gone in full to him and his younger colleagues. But there was no trace of resentment on his face, and in his subdued way he was thoroughly excited that the structure would prove of great benefit to biology.[3]

He was back in London only two days before he rang up to say that both he and Rosy found that their X-ray data strongly supported the double helix. They were quickly writing up their results and wanted to publish simultaneously with our announcement of the base pairs. *Nature* was the place for rapid publication, since if both Bragg and Randall strongly supported the manuscripts they might be published within a month of their receipt. However, there would not be only one paper from King's. Rosy and Gosling would report their results separately from Maurice and his collaborators.[4,5]

Rosy's instant acceptance of our model at first amazed me. I had feared that her sharp, stubborn mind, caught in her self-made antihelical trap, might dig up irrelevant results that would foster uncertainty about the correctness of the double helix. Nonetheless, like almost everyone else, she saw the appeal of the base pairs and accepted the fact that the structure was too pretty not to be true. Moreover, even before she learned of our proposal, the X-ray evidence had been forcing her more than she cared to admit toward a helical structure. The positioning of the backbone on the outside of the molecule was demanded by her evidence and, given the necessity to hydrogen-bond the bases together, the uniqueness of the A-T and G-C pairs was a fact she saw no reason to argue about.

At the same time, her fierce annoyance with Francis and me collapsed. Initially we were hesitant to discuss the double helix with her, fearing the

*See preliminary reaction
of Structure B on
17-2-53 p. 28*
A. K.

Structure B

23.2.53

Photograph 51 c

3·4 A arc   ~ 158·5 mm on projection

∴ 158·5 = 2R tan 2θ   where R is effective

specimen-film distance for projection

For d = 3·40 A,   θ = 13° 4'   tan 2θ = 0·491
calcd

R = $\frac{158.5}{2 \times 0.491}$ = 161·4 mm   2R = 322·8

[4] Franklin's notes shown here reveal that she began analyzing the B form in February 1953. (The annotation in upper left is by Aaron Klug. See also discussion on page 179.)

ROUGH   DRAFT   1.

A NOTE ON MOLECULAR CONFIGURATION IN SODIUM THYMONUCLEATE

Rosalind E. Franklin and R. G. Gosling

17/3/53.

Sodium thymonucleate fibres give two distinct types of X-ray diagram. The first, corresponding to a crystalline form obtained at about 75% relative humidity, has been described in detail elsewhere(.). At high humidities a new structure, showing a lower degree of order appears, and persists over a wide range of ambient humidity and water content. The water content of the fibres, which are crystalline at lower humidities, may vary from about 50% to several hundred per cent. of the dry weight in this structure. Other fibres which do not give crystalline structure at all, show this less ordered structure at much lower humidities. The diagram of this structure, which we have called structure B, shows in striking manner the features characteristic of helical structures ( ). Although this cannot be taken as proof that the structure is helical, other considerations make the existence of a helical structure highly probable.

[5] Unbeknown to Watson, Crick, and Wilkins, Franklin and Gosling had been preparing a manuscript summarizing their work and were quickly able to adapt it to the changed circumstances having seen Watson and Crick's model. This is a later version of the draft manuscript begun by Franklin and Gosling in early March 1953.

testiness of our previous encounters. But Francis noticed her changed attitude when he was in London to talk with Maurice about details of the X-ray pictures. Thinking that Rosy wanted nothing to do with him, he spoke largely to Maurice, until he slowly perceived that Rosy wanted his crystallographic advice and was prepared to exchange unconcealed hostility for conversation between equals. With obvious pleasure Rosy showed Francis her data, and for the first time he was able to see how foolproof was her assertion that the sugar phosphate backbone was on the outside of the molecule. Her past uncompromising statements on this matter thus reflected first-rate science, not the outpourings of a misguided feminist.

Obviously affecting Rosy's transformation was her appreciation that our past hooting about model building represented a serious approach to science, not the easy resort of slackers who wanted to avoid the hard work necessitated by an honest scientific career. It also became apparent to us that Rosy's difficulties with Maurice and Randall were connected with her understandable need for being equal to the people she worked with. Soon after her entry into the King's lab, she had rebelled against its hierarchical character, taking offense because her first-rate crystallographic ability was not given formal recognition.

Two letters from Pasadena that week brought the news that Pauling was still way off base. The first came from Delbrück, saying that Linus had just given a seminar during which he described a modification of his DNA structure. Most uncharacteristically, the manuscript he had sent to Cambridge had been published before his collaborator, R. B. Corey, could accurately measure the interatomic distances. When this was finally done, they found several unacceptable contacts that could not be overcome by minor jiggling.[6] Pauling's model was thus also impossible on straightforward stereochemical grounds. He hoped, however, to save the situation by a modification suggested by his colleague Verner Schomaker.[7] In the revised form the phosphate atoms were twisted 45 degrees, thereby allowing a different group of oxygen atoms to form a hydrogen bond. After Linus' talk, Delbrück told Schomaker he was not convinced that Linus was right,

[6] Pauling had written to Peter on February 18, 1953, telling him that "I am checking over the nucleic acid structure again, trying to refine the parameters a bit. I think that the original parameters are not exactly right. It is evident that the structure involves a tight squeeze for nearly all the atoms."

Robert Corey.

Verner Schomaker.

[7] Robert Corey was Pauling's right-hand man, a rigorous experimentalist who tested Pauling's ideas, carrying out the necessary X-ray analyses. Corey also developed CPK space-filling atomic models.

Verner Schomaker was a chemist at Caltech, specializing in electron and X-ray diffraction, and noted for his wide-ranging interests and prodigious intellect.

for he had just received my note saying that I had a new idea for the DNA structure.

Delbrück's comments were passed on immediately to Pauling, who quickly wrote off a letter to me. The first part betrayed nervousness—it did not come to the point, but conveyed an invitation to participate in a meeting on proteins to which he had decided to add a section on nucleic acids. Then he came out and asked for the details of the beautiful new structure I had written Delbrück about. Reading his letter, I drew a deep breath, for I realized that Delbrück did not know of the complementary double helix at the time of Linus' talk. Instead, he was referring to the like-with-like idea. Fortunately, by the time my letter reached Cal Tech the base pairs had fallen out. If they had not, I would have been in the dreadful position of having to inform Delbrück and Pauling that I had impetuously written of an idea which was only twelve hours old and lived only twenty-four before it was dead.

Todd made his official visit late in the week, coming over from the chemical laboratory with several younger colleagues. Francis' quick verbal

[8]This picture was taken by Antony Barrington Brown around the time the structure of DNA was published. As a student of Natural Sciences at Gonville and Caius, Barrington Brown spent more time working on the student newspaper (*Varsity*) than on his studies, and as picture editor once fired a young photographer named Anthony Armstrong-Jones (later Lord Snowdon). Soon after graduating, Barrington Brown set himself up as a professional photographer and it was in this capacity that he was asked to photograph the double helix model by a friend who hoped to sell a story about DNA to *Time* magazine. But *Time* chose not to publish the story and returned the negatives with half a guinea as compensation. It is believed that the now famous photo was in fact not published until it appeared in *The Double Helix* in 1968. On the same day Barrington Brown also took the photo of Watson and Crick having morning tea in their office (Chapter 29) and the photo of Crick fiddling with the DNA model in Chapter 27.

Watson and Crick in front of the DNA model.

tour through the structure and its implications lost none of its zest for having been given several times each day for the past week. The pitch of his excitement was rising each day, and generally, whenever Jerry or I heard the voice of Francis shepherding in some new faces, we left our office until the new converts were let out and some traces of orderly work could resume.[8] Todd was a different matter, for I wanted to hear him tell Bragg that we had correctly followed his advice on the chemistry of the sugar-phosphate backbone. Todd also went along with the keto configurations, saying that his organic-chemist friends had drawn enol groups for purely arbitrary reasons. Then he went off, after congratulating me and Francis for our excellent chemical work.[9]

Soon I left Cambridge to spend a week in Paris. A trip to Paris to be with Boris and Harriett Ephrussi had been arranged some weeks earlier. Since the main part of our work seemed finished, I saw no reason to postpone a visit which now had the bonus of letting me be the first to tell Ephrussi's and Lwoff's labs about the double helix. Francis, however, was not happy, telling me that a week was far too long to abandon work of such extreme significance. A call for seriousness, however, was not to my liking—especially when John had just shown Francis and me a letter from Chargaff in which we were mentioned. A postscript asked for information on what his scientific clowns were up to.[10]

[9] Surely Watson's reference to Todd complimenting them on their "chemical" work is ironic! Todd did write admiringly in his autobiography: "When I saw the Watson–Crick model that day in their laboratory, I at once recognised that, by a brilliant imaginative jump, they had not only solved the basic problem of a self-replicating molecule, but had thereby opened the way to a new world in genetics."

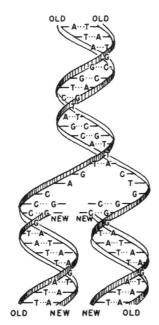

The manner envisaged for DNA replication, given the complementary nature of the base sequences in the two chains.

[10] As shown in this diagram, the two strands of the double helix must separate during DNA replication. However, as Delbrück explained to Watson, this leads to problems: "I am willing to bet that the plectonemic coiling of the chains in your model is radically wrong, because... the difficulties of untangling the chains do seem, after all, insuperable to me..." (May 12, 1953). J. B. S. Haldane was more pragmatic when he saw the DNA model at the Royal Society *conversazione* in July 1953. Ray Gosling describes Haldane as "...puffing on this foul Woodbine [a cheap English cigarette], and he looked at it [the model] for a long time, and then he said, 'So what you want is an "untwiddle-ase" enzyme.'" This, topoisomerase, was duly discovered 18 years later.

Chapter 29

Pauling first heard about the double helix from Delbrück. At the bottom of the letter that broke the news of the complementary chains, I had asked that he not tell Linus.[1] I was still slightly afraid something would go wrong and did not want Pauling to think about hydrogen-bonded base pairs until we had a few more days to digest our position. My request, however, was ignored. Delbrück wanted to tell everyone in his lab and knew that within hours the gossip would travel from his lab in biology to their friends working under Linus. Also, Pauling had made him promise to let him know the minute he heard from me. Then there was the even more important consideration that Delbrück hated any form of secrecy in scientific matters and did not want to keep Pauling in suspense any longer.

Pauling's reaction was one of genuine thrill, as was Delbrück's. In almost any other situation Pauling would have fought for the good points of his idea. The overwhelming biological merits of a self-complementary DNA molecule made him effectively concede the race. He did want, however, to see the evidence from King's before he considered the matter a closed book. This he hoped would be possible three weeks hence, when he would come to Brussels for a Solvay meeting on proteins in the second week of April.[2]

Linus Pauling's notebook from the Solvay conference.

[3] Watson described these pleasing results in a letter to Delbrück (March 22, 1953): "Concerning our assumption as to the equivalence of (adenine and thymine) and (guanine and cytosine) several days ago I ran into Wyatt at the Institut Pasteur. He tells me that the more he refines the analysis of the bases the clearer he finds the 1 to 1 equivalence. This 1 to 1 ratio also holds for 5 methyl-hydroxy cytosine which after more careful analysis seems to be equal to guanine."

That Pauling was in the know came out in a letter from Delbrück, arriving just after I returned from Paris on March 18. By then we didn't mind, for the evidence favoring the base pairs was steadily mounting. A key piece of information was picked up at the Institut Pasteur. There I ran into Gerry Wyatt, a Canadian biochemist who knew much about the base ratios of DNA. He had just analyzed the DNA from the T2, T4, and T6 group of phages. For the past two years this DNA was said to have the strange property of lacking cytosine, a feature obviously impossible for our model. But Wyatt now said that he, together with Seymour Cohen and Al Hershey, had evidence that these phages contained a modified type of cytosine called 5-hydroxy-methyl cytosine. Most important, its amount equaled the amount of guanine. This beautifully supported the double helix, since 5-hydroxy-methyl cytosine should hydrogen-bond like cytosine. Also pleasing was the great accuracy of the data, which illustrated better than any previous analytical work the equality of adenine with thymine and guanine with cytosine.[3,4]

While I was away Francis had taken up the structure of the DNA molecule in the A form. Previous work in Maurice's lab had shown that crystalline

The Bases of the Nucleic Acids of some Bacterial and Animal Viruses: the Occurrence of 5-Hydroxymethylcytosine

By G. R. WYATT
Laboratory of Insect Pathology, Sault Ste Marie, Ontario, Canada

and S. S. COHEN
Department of Pediatrics, Children's Hospital of Philadelphia and Department of Physiological Chemistry, University of Pennsylvania, Philadelphia, Pennsylvania

(*Received* 11 *April* 1953)

Recent studies on the multiplication of viruses have directed attention increasingly toward their nucleic acids. Hershey & Chase (1952) have shown that most, if not all, of the sulphur-containing protein of coliphage T2, which appears to be present in the outer shell of the virus, does not enter the infected cell. However, deoxyribonucleic acid (DNA), apparently organized within the virus, is in some way transferred to the host cell, and appears, therefore, to participate more intimately in the transmission of genetic properties. On infection of *Escherichia coli* with bacteriophage T2, T4 or T6, there is immediate cessation of synthesis of ribonucleic acid (RNA) and net synthesis of DNA is detectable in about 10 min. (Cohen, 1947, 1951). A similar apparent redirection of DNA synthesis during virus multiplication is characteristic of certain induced lysogenic systems, but in this case synthesis of RNA

Wyatt's analysis of base composition.

mum 274 mμ., as yet unidentified. When combined, these products had an absorption spectrum close to that of deoxycytidylic acid. The conclusions drawn from these studies, however, are not altered by the substitution of hydroxymethylcytosine for cytosine.

Marshak (1951) missed hydroxymethylcytosine because of his use of perchloric acid for hydrolysis along with a chromatogram solvent system in which it happens to migrate together with guanine. This accounts for the anomalous absorption spectrum for guanine which he reported.

In spite of the considerable evidence that DNA may play a specific role in the transmission of hereditary characters, we were unable to demonstrate any difference in the composition of the DNA of the r and r$^+$ mutants of phages T 2, T 4 and T 6. This confirms the inference drawn from similar

ments in technique have resulted in bringing the observed ratios successively closer to unity. One is tempted to speculate that regular structural association of nucleotides of adenine with those of thymine and of guanine with those of cytosine (or its derivatives) in the DNA molecule requires that they be equal in number. There is as yet, however, no direct evidence for such a theory.*

The occurrence of 5-hydroxymethylcytosine as a major constituent of the nucleic acid of a virus, none of which could be found in the host cells, presents problems of fundamental importance for the chemistry of virus production. Although discussion must at present remain largely speculative, certain possibilities may be pointed out.

We are concerned with the following pyrimidine bases:

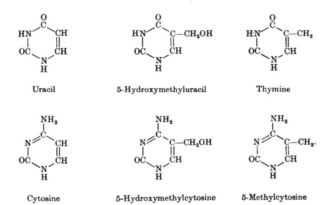

Uracil   5-Hydroxymethyluracil   Thymine

Cytosine   5-Hydroxymethylcytosine   5-Methylcytosine

analyses on a number of insect viruses (Wyatt, 1952b) that genetic difference is not necessarily accompanied by a detectable quantitative difference in DNA composition.

A common pattern has been noted in the composition of DNA from many sources: the molar ratios (adenine)/(thymine) and (guanine)/(cytosine + 5-methylcytosine) are relatively constant and close to unity (Chargaff, 1951; Wyatt, 1952a). The same regularities are seen to be valid with DNA from phage T 5 and from vaccinia virus, and also with DNA of phages T 2, T 4 and T 6 except that here cytosine is replaced by 5-hydroxymethylcytosine. Whether these near-unity ratios actually signify equal numbers of the corresponding nucleotides in the molecule is as yet uncertain. The present studies, however, have served to emphasize how quantitative errors can result from small differences in experimental conditions and purity of materials, and it is our experience that successive improve-

The metabolic pathways for pyrimidines appear generally to involve their ribosides and deoxyribosides rather than the free bases, and preliminary experiments by one of us (S. S. C.) indicate that this probably is the case in *Esch. coli*. In the rat, Reichard & Estborn (1951) have demonstrated that deoxycytidine can be utilized for production of thymidine, but not vice versa. Elwyn & Sprinson (1950) have implicated the β-carbon of serine as a source of the 5-methyl group of thymine, which is evidently synthesized by methylation of a preformed pyrimidine ring. Since serine cleaves to formaldehyde, we may question whether methyl-group synthesis from serine may not involve an initial hydroxymethylation followed by reduction. If this is so, 5-hydroxymethylpyrimidines (or their deoxyribosides) could be normal metabolites, inter-

* Since this was written, a structure for DNA involving such specific pairing of nucleotides has been proposed by Watson & Crick (1953).

[4]This paper was submitted for publication (April 11) before the DNA structure was published (April 25), but knowledge of the basic model is indicated by the footnote shown here.

A-form DNA fibers increase in length when they take up water and go over into the B form. Francis guessed that the more compact A form was achieved by tilting the base pairs, thereby decreasing the translational distance of a base pair along the fiber axis to about 2.6 Å. He thus set about building a model with tilted bases. Though this proved more difficult to fit together than the more open B structure, a satisfactory A model awaited me upon my return.

In the next week the first drafts of our *Nature* paper got handed out and two were sent down to London for comments from Maurice and Rosy. They had no real objections except for wanting us to mention that Fraser in their lab had considered hydrogen-bonded bases prior to our work. His schemes, until then unknown to us in detail, always dealt with groups of three bases, hydrogen-bonded in the middle, many of which we now knew to be in the wrong tautomeric forms. Thus his idea did not seem worth resurrecting only to be quickly buried. However, when Maurice sounded upset at our objection, we added the necessary reference. Both Rosy's and Maurice's papers covered roughly the same ground and in each case interpreted their results in terms of the base pairs. For a while Francis wanted to expand our note to write at length about the biological implications. But finally he saw the point to a short remark and composed the sentence: "It has not escaped our notice that the specific pairing we have postulated immediately suggests a possible copying mechanism for the genetic material."[5]

Sir Lawrence was shown the paper in its nearly final form. After suggesting a minor stylistic alteration, he enthusiastically expressed his willingness to post it to *Nature* with a strong covering letter. The solution to the structure was bringing genuine happiness to Bragg. That the result came out of the Cavendish and not Pasadena was obviously a factor. More important was the unexpectedly marvelous nature of the answer, and the fact that the X-ray method he had developed forty years before was at the heart of a profound insight into the nature of life itself.

The final version was ready to be typed on the last weekend of March. Our Cavendish typist was not on hand, and the brief job was given to my sister. There was no problem persuading her to spend a Saturday afternoon this

[5] On this and the following pages, we deal with the correspondence on publishing the paper. Thus, a draft letter from Crick to Wilkins on March 17:

"Dear Maurice,

I enclose a draft of our letter. As this has not been seen by Bragg I would be grateful if you did not show it to anyone else. The object of sending it to you at this stage is to obtain your approval of two points: a) the reference number 8 to your unpublished work. b) the acknowledgement. If you would like either of these rewritten, please let us know. If we don't hear from you within a day or so we shall assume that you have no objection to their present form. Jim has gone to Paris, lucky dog.

Yours"

The timings of Watson's trip to Paris and the sending of their first draft to London seem slightly off here. From Wilkins' reply to Crick, March 18:

"I think you're a couple of old rogues but you may well have something. I like the idea. Thanks for the MSS. I was a bit peeved because I was convinced that 1:1 purine pyrimidine ratio was significant and had a 4 planar group sketch and was going to look into it and as I was back again on helical schemes I might, given a little time, have got it. But there is *no good grousing*—I think it's a very exciting notion and who the hell got it isn't what matters.

…We should like to publish a brief note with a picture showing the general helical case…

Just heard this moment of a new entrant in the helical rat-race. R.F. and G. have served up a rehash of our ideas of 12 months ago. It seems they should publish something too (they have it all written). So at least 3 short articles in *Nature*. As one rat to another, good racing."

Correspondence about the manuscripts continued. Wilkins wrote to Crick on March 23 expressing as vividly as ever his feelings. "I am so browned off with the whole madhouse I don't really care much what happens…"

He discusses a number of issues. First, coordinating the papers ("It looks as though the only thing is to send Rosy's and my letters as they are and hope the editor doesn't spot the duplication" and ending with a "PS. Raymond and Rosie have your thing so everyone will have seen everybody else's"). Second, the imminent visit of Pauling to both Cambridge and London ("If Rosy wants to see Pauling, what the hell can we do about it?…Now Raymond wants to see Pauling too! To hell with it all."). And third, he defends his position on including reference to Fraser's model in Watson and Crick's paper:

"I feel your remarks about Bruce's model, in your note, not in very good style. Why be bitter about it? I refer to 'It is stated that the bases…on it.' I suggest (In his model the phosphates are on the outside and the bases are hydrogen bonded in planar groups inside. The model has many (or serious) weaknesses which we will not discuss here.) and put it all in brackets."

BIOPHYSICS RESEARCH UNIT,
KING'S COLLEGE,
STRAND,
LONDON, W.C.2.
TELEPHONE : TEMPLE BAR 5651

Mar.

Dear Francis

It looks as though the only thing is to send Rosy's & my letters as they are & hope the Editor doesn't spot the duplication. I am so browned off with the whole madhouse I don't really care much what happens. If Rosy wants to see Pauling what the hell can we do about it? If we suggested it would be nicer if she didn't that would only encourage her to do so. why is everybody so terribly interested in seeing Pauling?

If you like to put in a good word for me for a trip to Pasadena OK. We will post a copy of Rosy's thing to you tomorrow. I don't see why we have to have a meeting

I feel your remarks about Bruce's model, in your note,

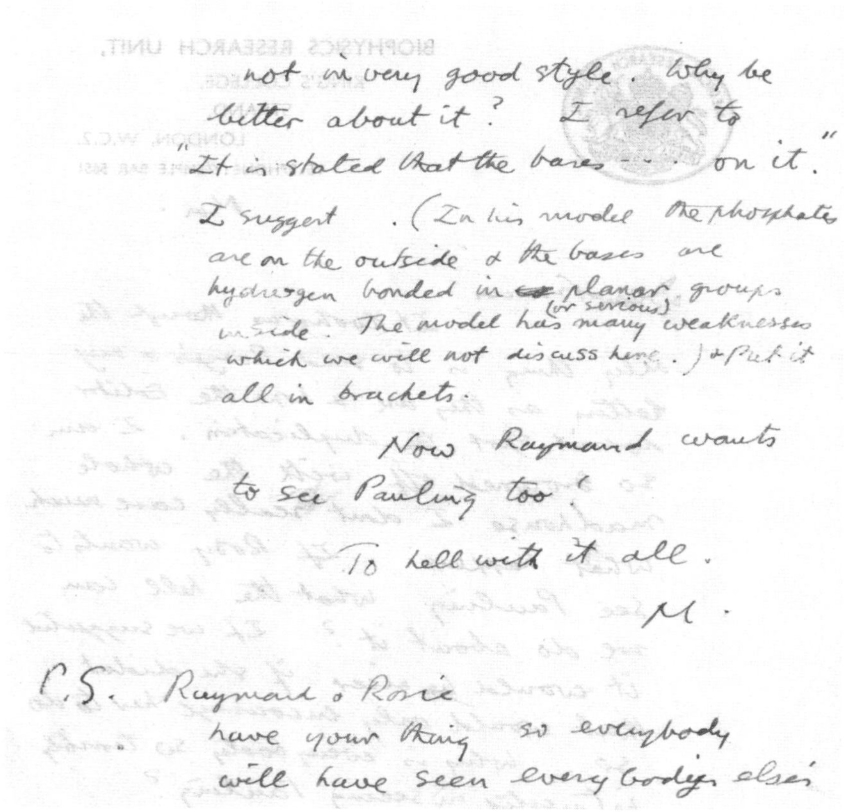

not in very good style. Why be bitter about it? I refer to "It is stated that the bases --- on it."

I suggest . (In his model the phosphates are on the outside & the bases are hydrogen bonded in ~~as~~ planar groups (or serious) inside. The model has many weaknesses which we will not discuss here.) & Put it all in brackets.

Now Raymond wants to see Pauling too!

To hell with it all.

M.

C.S. Raymond & Rosie have your thing so everybody will have seen everybody else's

way, for we told her that she was participating in perhaps the most famous event in biology since Darwin's book. Francis and I stood over her as she typed the nine-hundred-word article that began, "We wish to suggest a structure for the salt of deoxyribose nucleic acid (DNA). This structure has novel features which are of considerable biological interest." On Tuesday the manuscript was sent up to Bragg's office and on Wednesday, April 2, went off to the editors of *Nature*.[6]

Linus arrived in Cambridge on Friday night. On his way to Brussels for the Solvay meeting, he stopped off both to see Peter and to look at the model.

Gerard Pomerat.

[6] Gerard Pomerat, assistant director of the natural science program at the Rockefeller Foundation, was visiting the Cavendish on April 1, 1953. His diary provides the only contemporary independent account of the atmosphere there between discovering the structure and publishing the paper.

"There was at the Cavendish today a great air of excitement…They believe they have really got the structure of nucleic acid from a crystallographic rather than a chemical standpoint. Their clue came out of the beautiful X-ray diagrams produced in Randall's lab and some of the work which had meanwhile been going on at Cambridge. They are just putting the finishing touches on a huge model about six feet tall…[The two chaps] are J. D. Watson and F. H. C. Crick…Both young men are somewhat mad hatters who bubble over about their new structure in characteristic Cambridge style and it is hard to realize that one of them is an American…[They] are certainly not lacking, however, in either enthusiasm or ability."

Unthinkingly Peter arranged for him to stay at Pop's. Soon we found that he would have preferred a hotel. The presence of foreign girls at breakfast did not compensate for the lack of hot water in his room. Saturday morning Peter brought him into the office, where, after greeting Jerry with Cal Tech news, he set about examining the model. Though he still wanted to see the quantitative measurements of the King's lab, we supported our argument by showing him a copy of Rosy's original B photograph. All the right cards were in our hands and so, gracefully, he gave his opinion that we had the answer.[7]

Bragg then came in to get Linus so that he could take him and Peter to his house for lunch. That night both Paulings, together with Elizabeth and me, had dinner with the Cricks at Portugal Place. Francis, perhaps because of Linus' presence, was mildly muted and let Linus be charming to my sister and Odile. Though we drank a fair amount of burgundy, the conversation never got animated and I felt that Pauling would rather talk to me, clearly an

Morning tea in the Cavendish just after publication of the manuscript on the double helix.

CLUB DE LA FONDATION UNIVERSITAIRE

TÉLÉPHONES { 11.81.00 (4 LIGNES)
{ 12.24.22 (PERMANENT ET LE DIMANCHE)
CHÈQUES POSTAUX N° 1039.46
ADRESSE TÉLÉGRAPHIQUE : " FONDUNI-BRUXELLES "

BRUXELLES, LE 6 April 1953
11, RUE D'EGMONT

Dearest little love:

I've just arrived, safely, in Brussels, and am in my room. No one else seems to be here. Synge and Adair were in the London bus terminal, and Peter and I had coffee with them in the buffet, about noon. They came on the plane with me (S + A) but are in a hotel, rather than this club. I shall go for a walk now, and have dinner, and go to bed. The town seems to be very quiet — it is Easter Monday, a bank holiday in England.

I saw Dorothy and Thomas — they passed through Cambridge on the way to visit D's mother.

Peter is in fine shape — his face not yet clear, but perhaps a bit better. He is cutting a wisdom tooth. He flies tomorrow to Paris. He liked the stuffed dates and cookies. I overlooked his vest. Anyway, he has a bright red one, with brass buttons — they are popular with the boys.

We had lunch with the Braggs, & their two daughters. The talk was much about Trellin. The Braggs are looking forward to their trip to California.

I dined with Bragg & Roughton last night in Trinity College. I didn't see the Todds, nor the Rothschilds & Tylers — they were in Rushbrooke.

The flight from London was nice, but a bit cloudy & a bit rough. Our bus from the airport met a 4-engined plane coming toward us on the road. Our driver turned the bus around, shot up the road ahead of the plane (which took up the whole road) and then off to one side. This is an unusual hazard.

I haven't yet learned what our schedule is for the week, nor who will be here (some I know about). I am already pretty lonesome, after five days.

I have seen the King's College nucleic acid pictures, and talked with Watson and Crick, and I think that our structure is probably wrong, and theirs right.

Much love from
Daddy

[7] Pauling's graceful acceptance that Watson and Crick were correct is revealed in letters he wrote in April. Shown is the charming letter to his wife Ava upon arriving in Belgium on April 6, 1953. The final paragraph reads:

"I have seen the King's College nucleic acid pictures, and talked with Watson and Crick, and I think that our structure is probably wrong, and theirs right."

And on April 20, he wrote to Delbrück:

"I am very deeply impressed by the Watson–Crick structure…While there is still a chance that their structure is wrong, I think that it is highly probable that it is right. It has very important implications, as you mention. It (sic) think that it is the most significant step forward that has been taken for a long time."

unfinished member of the younger generation, than to Francis. The talk did not last long, since Linus, still on California time, was becoming tired, and the party was over at midnight.

Elizabeth and I flew off the following afternoon to Paris, where Peter would join us the next day. Ten days hence she was sailing to the States on her way to Japan to marry an American she had known in college. These were to be our last days together, at least in the carefree spirit that had marked our escape from the Middle West and the American culture it was so easy to be ambivalent about. Monday morning we went over to the Faubourg St. Honoré for our last look at its elegance. There, peering in at a shop full of sleek umbrellas, I realized one should be her wedding present and we quickly had it. Afterwards she searched out a friend for tea while I walked back across the Seine to our hotel near the Palais du Luxembourg. Later that night with Peter we would celebrate my birthday. But now I was alone, looking at the long-haired girls near St. Germain des Prés and knowing they were not for me. I was twenty-five and too old to be unusual.[8]

[8] In a similar mood, Watson wrote to Delbrück (March 23, 1953): "I have a rather strange feeling about our DNA structure. If it is correct, we should obviously follow it up at a rapid rate. On the other hand, it will at the same time be difficult to avoid the desire to forget completely about nucleic acid and to concentrate on other aspects of life. This latter mood dominated me in Paris which as expected is by far the most interesting city I shall ever know."

Café at St. Germain des Prés, circa 1950.

Epilogue

[1] Herman Kalckar (1908–1991) moved first to the National Institutes of Health and spent the rest of his career in the U.S.

[2] Kendrew (1917–1997) became increasingly involved in science policy and was the leader in the establishment of the European Molecular Biology Organization and Laboratory.

[3] Perutz (1914–2002) remained at the Laboratory of Molecular Biology for the rest of his career. He frequently wrote reviews for *The New York Review of Books* and published collections of essays.

[4] Lawrence Bragg (1890–1971) set about reviving the fortunes of the Royal Institution, concentrating on education. He continued to make contributions to the determination of protein structure. In 1965, Bragg celebrated the 50th anniversary of the award of his Nobel Prize.

[5] Hugh Huxley spent two years at MIT before returning to the UK. Huxley joined the Laboratory of Molecular Biology, Cambridge, in 1962 and in 1987 he moved to Brandeis University where he is Professor Emeritus of Biology.

Virtually everybody mentioned in this book is alive and intellectually active. Herman Kalckar[1] has come to this country as professor of biochemistry at Harvard Medical School, while John Kendrew and Max Perutz both have remained in Cambridge, where they continue their X-ray work on proteins, for which they received the Nobel Prize in Chemistry in 1962.[2,3] Sir Lawrence Bragg retained his enthusiastic interest in protein structure when he moved in 1954 to London to become director of the Royal Institution.[4] Hugh Huxley, after spending several years in London, is back in Cambridge doing work on the mechanism of muscle contraction.[5] Francis Crick, after a year in Brooklyn, returned to Cambridge to work on the nature and operation of the genetic code, a field of which he has been the acknowledged world leader for the past decade.[6] Maurice Wilkins' work remained centered on DNA for some years until he and his collaborators established beyond any doubt that the essential features of the double helix were correct. After then making an important contribution to the structure of ribonucleic acid, he has changed the direction of his research to the organization and operation of nervous systems.[7] Peter Pauling now lives in London, teaching chemistry at University College. His father, recently retired from active teaching at Cal Tech, at present concentrates his scientific activity both on the structure of the atomic nucleus and on theoretical structural chemistry.[8] My sister, after being many years in the Orient, lives with her publisher husband and three children in Washington.

All of these people, should they desire, can indicate events and details they remember differently. But there is one unfortunate exception. In 1958, Rosalind Franklin died at the early age of thirty-seven. Since my initial impressions of her, both scientific and personal (as recorded in the early pages of this book), were often wrong, I want to say something here about her achievements. The X-ray work she did at King's is increasingly regarded as superb. The sort-

[6] Crick (1916–2004) turned his attention to embryonic development and then the organization of DNA in chromosomes. In 1977, he joined the Salk Institute studying memory and the nature of consciousness.

[7] Wilkins (1914–2006) pursued his interests in science and society. He became the first president of the British Society for Social Responsibility in Science, and was active in Pugwash and the Campaign for Nuclear Disarmament.

[8] Pauling (1901–1994) retired from Caltech in 1963 and became an advocate for taking large doses of vitamin C to prevent colds and cancer. In 1973, he founded the Institute of Orthomolecular Medicine to promote research in this field.

[9] Franklin was ranked at the highest level in her field. For example, she presented a paper at the elite CIBA Symposium on *The Nature of Viruses*, in 1957, the only woman among 34 participants, six of whom went on to win Nobel Prizes. In his closing remarks Sir Charles Harrington referred to the "…very beautiful papers read by Dr. Williams and Dr. Franklin."

ing out of the A and B forms, by itself, would have made her reputation; even better was her 1952 demonstration, using Patterson superposition methods, that the phosphate groups must be on the outside of the DNA molecule. Later, when she moved to Bernal's lab, she took up work on tobacco mosaic virus and quickly extended our qualitative ideas about helical construction into a precise quantitative picture, definitely establishing the essential helical parameters and locating the ribonucleic chain halfway out from the central axis.[9]

Because I was then teaching in the States, I did not see her as often as did Francis, to whom she frequently came for advice or when she had done something very pretty, to be sure he agreed with her reasoning. By then all traces of our early bickering were forgotten, and we both came to appreciate greatly her personal honesty and generosity, realizing years too late the struggles that the intelligent woman faces to be accepted by a scientific world which often regards women as mere diversions from serious thinking. Rosalind's exemplary courage and integrity were apparent to all when, knowing she was mortally ill, she did not complain but continued working on a high level until a few weeks before her death.

Taken at the International Union of Crystallography Symposium in Madrid on April 2, 1956, this photograph shows Franklin among her professional colleagues. From the left, Ann Cullis, Francis Crick, Don Caspar, Aaron Klug, Rosalind Franklin, Odile Crick, and John Kendrew.

The Nobel Prize

This is an abridged version of the chapter "Manners Appropriate for a Nobel Prize" in Watson's Avoid Boring People *(Alfred A. Knopf and Oxford University Press, 2007).*

[1] The Karolinska Institutet was founded by King Karl XIII in 1810 to train army surgeons. It is a group of 50 professors at the Karolinska, the Nobel Assembly, which awards the Nobel Prize in Physiology or Medicine.

Individuals nominated for Nobel Prizes are not supposed to know their names have been put forward. The Swedish Academy, which judges candidates and awards the prize, makes this policy very explicit on their nomination forms. Jacques Monod, however, could not keep secret from Francis Crick that a member of the Karolinska Institutet in Stockholm had asked him to nominate us in January for the 1962 Nobel Prize in Physiology or Medicine.[1] In turn, Francis, when visiting Harvard that February to give a lecture, let the cat out of the bag at a Chinese restaurant where we were having supper. But he told me we should say nothing to anyone, lest it get back to Sweden.

The obverse of the Nobel Medal.

[2] Charles Huggins was a cancer researcher and physician at the University of Chicago whose research on the effects of hormones on prostate and breast cancer led to the award of the Nobel Prize for Physiology or Medicine in 1966. Huggins shared it with Peyton Rous for the latter's discovery, some 55 years earlier, that some spontaneous chicken tumors were caused by a virus.

That we might someday get the Nobel Prize for finding the double helix had been bruited about ever since our discovery. Just before my mother died in 1957, she was told by Charles Huggins, then the University of Chicago's best-known physician-scientist, that I was certain to be so honored.[2] Though many were initially skeptical that DNA replication involved strand separation, this doubting chatter went silent after the 1958 Meselson-Stahl experiment demonstrated that very phenomenon. Certainly the Swedish Academy had no doubt as to the correctness of the double helix when they awarded Arthur Kornberg half of the 1959 Physiology or Medicine prize for experiments demonstrating enzymatic synthesis of DNA. When photographed shortly after learning of his Nobel, a beaming Kornberg held a copy of our demonstration DNA model in his hands.

Arthur Kornberg with a model of DNA in 1959.

As the October 18 date for announcing the year's Nobel in Physiology or Medicine approached, I was naturally jittery. Conceivably the responsible Swedish professors had requested more than one nomination, reflecting split opinions during preliminary caucusing. Nonetheless, as I went to bed the night before the prize announcement, I couldn't help fantasizing about being awakened by an early morning phone call from Sweden. Instead a nasty cold I'd caught awakened me prematurely, and I was depressed to realize at once that no word had come from Stockholm. I remained shivering under my electric blanket, not wanting to get up when the telephone rang at 8:15 A.M. Rushing into the next room, I happily heard a Swedish newspaper reporter's voice tell me that Francis Crick, Maurice Wilkins, and I had won the Nobel Prize for Physiology or Medicine. Asked how I felt, all I could say was, "Wonderful!"

First I phoned Dad and then my sister, inviting each to accompany me to Stockholm. Soon after, my telephone began to buzz with congratulatory messages from friends who had already heard the news on the morning broadcasts. There were also calls coming from reporters, but I told them to try me at Harvard after I'd given my morning virus class. I felt no need to rush through breakfast with Dad, so the class hour was almost half over when I walked in to find an overflowing crowd of students and friends anticipating my arrival. The words *Dr. Watson has just won the Nobel Prize* were on the blackboard.

The crowd clearly did not want a virus lecture, so I spoke about feeling the same elation when we first saw how base pairs fitted so perfectly into a DNA double helix, and how pleased I was that Maurice Wilkins was sharing the prize. It was his crystalline A-form X-ray photograph that had told us there was a highly regular DNA structure out there to find. If Linus Pauling's ill-conceived structure had not gotten Francis and me back into the DNA game, Maurice, keen to resume work on DNA the moment Rosalind Franklin moved over to Birkbeck College, might by himself have been the first to see the double helix. He was temporarily in the States when the prize story broke, and held his press conference next to a big DNA model at the Sloan-Kettering Institute. The long-standing rule that a Nobel Prize can be

Nobel Winner Holds Regular Class

A newspaper clipping of Watson talking to his Harvard class on the morning of the announcement of the Nobel Prize.

shared by at most three individuals would have created an awkward if not insolvable dilemma had Rosalind Franklin still been alive. But having been tragically diagnosed with ovarian cancer less than four years after the double helix was found, she'd died in the spring of 1958.

After class ended, I soon found myself with a champagne glass in hand and talking to reporters from the Associated Press, United Press International, the *Boston Globe*, and *Boston Traveler*. Their stories were picked up by most papers across the country, clippings from which came to me through the Harvard news office. Often they were accompanied by AP photos showing me in front of my class or holding the hand-size demonstration model of the double helix built at the Cavendish back in 1953. Able to afford the luxury of modesty, I tried to downplay potential practical applications, saying that a cure for cancer was not an obvious consequence of our work. And with my stuffy head and hoarse voice quite apparent, I emphasized that we had not done away with the common cold. This became the quotation of the day in the October 19 *New York Times*. When asked how I would spend the money, I said possibly on a house and most certainly not on hobbies such as stamp collecting. To the question as to whether our work might lead to genetically improving humans, I answered, "If you want to have an intelligent child, you should have an intelligent wife."

A hastily arranged evening blast at Paul and Helga Doty's Kirkland Place house allowed my Cambridge friends to toast my good fortune.[3] Earlier I had talked by phone to Francis Crick, no less elated in the other Cambridge.[4] Most of some eighty congratulatory telegrams arrived over the next

A press conference at Harvard, October 18, 1962.

The receipt for seven cases of Mercier champagne for the party at the Dotys'.

[3] *The receipt for seven cases of Mercier champagne for the party at the Dotys'.*

```
My dear Jim,

        It was nice of you to ring up on the 18th.  I'm sorry if
I was incoherent, but there was so much noise I could hardly
hear what you said.  I hear you had a good party the next day
and also that you plan to spend the prize on women!
```

[4] *Crick wrote to Watson on October 30, 1962, explaining that the celebratory party in Cambridge had gone well. In the same letter, Crick discusses the Nobel Lectures, proposing that Wilkins deals with the structure, Watson discusses RNA and Crick covers the genetic code.*

[5] Nathan M. Pusey was President of Harvard University from 1953 to 1971. He defended the faculty from the attacks of Joseph McCarthy but his tenure was marred by his actions during a student occupation of University Hall, the administrative building, in 1969. He called in state and local police in riot gear who used tear gas to clear the building. It was Pusey who vetoed Harvard University Press publishing *The Double Helix* (see Appendix 4).

[6] Henry Stuart Hughes was an historian at Harvard who ran as an independent for the unexpired Senate seat of John F. Kennedy against the President's youngest brother Edward M. Kennedy. Hughes was pro-disarmament, which was a liability following the Cuban missile crisis when the power of the U.S. military forced the U.S.S.R. to back down. Hughes was defeated decisively by Ted Kennedy.

two days, while the next week brought some two hundred letters I would eventually have to acknowledge. As he was then briefly hospitalized, Lawrence Bragg, our old boss at the Cavendish, had his secretary write of his delight. Unique in addressing me as "Mr. Watson," President Pusey wrote: "It seems almost superfluous to add my congratulations to the many friendly messages you will be receiving."[5] And I had to wonder whether I had been mistaken in backing Harvard's Stuart Hughes for the Senate when not he but Edward M. Kennedy took the time to write, "Your contribution is one of the most exciting scientific achievements of our time."[6]

There were also the inevitable letters expressing not congratulations but the writer's personal hobbyhorse. One from a Palm Beach man, for instance, declared that marriages between cousins are the cause of all the great evils that have afflicted mankind. Here I thought better of writing back to ask whether there had been any such marriages among his ancestors. More poignant, but strange nonetheless, was a letter from a seventeen-year-old Samoan girl in Pago Pago who, after thanking the Lord for his love and kindness, introduced herself as Vaisima T. W. Watson. She hoped that I was related to her father, Thomas Willis Watson, a U.S. Marine supply sergeant during World War II. Her mother had never heard from him following his return to the States. In my reply, I pointed out that Watson was a common name, with hundreds of entries in the Boston phone book alone.

Soon the itinerary of my forthcoming Nobel week, in broad outline at least, was sent to me from Stockholm. I would be housed in the Grand Hotel with my guests. My personal expenses there would be paid by the Nobel Foundation, which would also cover the food and lodging of a wife and any children. As getting there would be my responsibility, the Nobel Foundation would advance some of my prize money for airfare. The prize presentations were to take place at the Stockholm concert hall according to custom on December 10, the date on which Alfred Nobel died in San Remo, Italy, in 1896 at the age of sixty-three. I was expected to arrive several days earlier for two receptions, the first given by the Karolinska Institutet for its winners, and the second by the Nobel Foundation for all laureates except for those in

Peace, who always receive their prizes in Oslo from the Norwegian king. At the prize ceremony and at the banquet the following night at the Royal Palace, I was to be dressed in white tie and tails. Meeting me at the airport was to be a junior member of the Foreign Ministry, who would accompany me to all official functions and see me off at my departure.

Making it an even more meaningful occasion was the awarding of the year's chemistry prize to John Kendrew and Max Perutz for their respective elucidations of the three-dimensional structures of the proteins myoglobin and hemoglobin. Never before in Nobel history had one year's prizes in biology and chemistry gone to scientists working in the same university laboratory.[7] The announcement of John and Max's prize came several days after ours was announced, on the same day the physics prize was awarded to the Russian theoretical physicist Lev Landau for his pioneering studies on liquid helium. Unfortunately, because of a ghastly automobile crash that had recently caused him severe brain damage, he would not be joining us in Stockholm. After the double helix was found, the Russian-born physicist George Gamow had much raised my ego by saying that I reminded him of the young Landau.[8] Last to be announced was the literature prize, awarded to the novelist John Steinbeck, who was to deliver his Nobel address at the large banquet in the Stockholm city hall following the prize ceremonies.

John Kendrew.   *Max Perutz.*

[7] As shown above, at the Laboratory for Molecular Biology, John Kendrew and Max Perutz celebrate the announcement of the Nobel Prize for Chemistry.

[8] George Gamow, wearing the RNA Tie Club tie, at Cold Spring Harbor Laboratory in 1963. Gamow was a physicist whose most famous contribution was to cosmology, the Big Bang Theory. Following publication of the double helix papers, Gamow was one of the first to tackle the problem of the genetic code. Gamow wrote a series of popular books on science, featuring a fictional character named Mr. C. G. H. Tompkins. Tompkins was also Gamow's co-author on a paper submitted to the *Proceedings of the National Academy of Sciences*. The National Academy of Sciences was not amused and returned the manuscript.

Lev Landau.   *George Gamow.*

For several days, I eagerly anticipated a November 1 state dinner at the White House to which I received a last-minute invitation. Though the event was intended to honor the grand duchess of Luxembourg, I was more keen to see America's royal couple in action. Only six months had passed since they had elegantly honored the nation's 1961 Nobel Prize winners, so I now thought the occasion might find me seated beside Jackie. All such thoughts, however, were abruptly interrupted by the Cuban missile crisis. JFK's speech to the nation on Monday, October 20, was not one to be listened to alone. Nervously, I went to the Doty home to watch it on their relatively big TV screen. Even before the speech was over I knew the gravity of the situation was such that a politically unnecessary state dinner was bound to be cancelled. From then on the president's attention would necessarily be focused on whether the Soviets would challenge the American blockade of Cuba, in which case the prospect of nuclear war seemed very real indeed.

Paul Doty, Watson's mentor and supporter at Harvard.

Over the next several days, I had to wonder whether a month hence I would, in fact, be going to Stockholm. The Soviets might very well set up their own blockade of Berlin. Happily, less than a week passed before Nikita Khrushchev backed down. By then it was too late to reschedule the dinner for the grand duchess. The White House, however, kept me in view and invited me to a December luncheon for the president of Chile. But the thrill that came of seeing the White House envelope vanished when I opened it and saw that the date overlapped with Nobel week. I continued hoping that there might be a place for me at still another White House affair. But by the new calendar year I was no longer a celebrity of the moment.

A visit to the University of Chicago had been arranged some months before the Nobel Prize announcement. Suddenly it became a media event, with visits to my former grammar school and high school hastily arranged. Also making a return visit to Horace Mann Grammar School that day was Greta

Watson talks to the pupils of the Horace Mann Grammar School.

Brown, the principal when I was there between the ages of five and thirteen. Earlier she had penned me a warm letter recalling my bird-watching days and regretting that my very well-liked mother had not lived to enjoy my triumph. The school auditorium was crammed as I spoke from the stage, gazing once again upon its handsome big WPA murals. The next day, in the *Chicago Daily News* a nearly whole page spread was headlined "The Return of a Hero" and quoted a teacher remembering me as "very short in stature but with a very eager mind." Later at South Shore High School I spoke to an even larger audience including my former biology teacher, Dorothy Lee, who much encouraged me during my sophomore year.

The next day I flew on to San Francisco and went down to Stanford to talk science. Then I traveled across the bay to Berkeley, where I stayed with Don and Bonnie Glaser. Two years before, Don had won the Nobel Prize in Physics for his invention of the bubble chamber, causing them to advance their wedding date so that they could fly off together as man and wife to

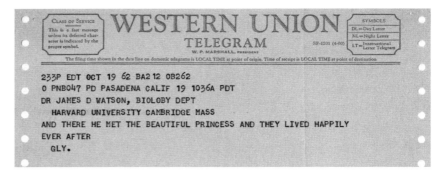

[9] *The congratulatory telegram from Richard Feynman. It is signed "Gly" for glycine, Feynman's nickname in the RNA Tie Club.*

Sweden. In her note of congratulations, Bonnie encouraged me to set my sights on a Swedish princess, suggesting Desiree for both her poise and beauty, as well as having more to say than her two older sisters. So I told them about a telegram from my Caltech friend, the physicist Dick Feynman, wherein he proposed the same scenario with even more irony: "And there he met the beautiful princess and they lived happily ever after."[9]

My forthcoming Nobel address soon preoccupied me at Harvard. Maurice was to give his talk on his King's College lab work confirming the double helix; Francis would focus on the genetic code; and I would talk about the involvement of RNA in protein synthesis. Happily, my Harvard science of the past five years was equal to a Nobel lecture. By then I had bought the necessary white-tie outfit at the Cambridge branch of J. Press, whose first shop in New Haven had long been purveyor par excellence of preppy clothing to Yale's undergraduates. Soon after coming to Harvard, I had begun getting my suits at their Mt. Auburn Street store, finding their clothes to be among the few available that fit my still-skinny frame. Perhaps sensing my high spirits, the salesman easily persuaded me also to purchase for the august occasion a black cloth coat with a fur collar.

Early on the afternoon of December 4 my sister joined Dad and me in New York for our Scandinavian Airlines flight. Our plans were to stop over

Watson, his father and sister pose for a Scandinavian Airline Systems (SAS) publicity photograph.

for two nights in Copenhagen to see friends Betty and I had made when we lived there in the early 1950s. But after crossing the Atlantic, the pilots discovered that Copenhagen was fogged in. So we found ourselves in Stockholm two days earlier than expected. Bypassing customs as if we were a diplomatic delegation, we were whisked by limousine to the storied Grand Hotel, built in 1874, across from the Royal Palace, onto which looked my room, among the finest in the house. I soon joined my sister and Dad for a herring-heavy smorgasbord, where we lunched with Kai Falkman, the young Swedish diplomat who would accompany us to all our Nobel week engagements.[10] Kai told us that the youngest of the four Swedish princesses, Christina, wanted to spend a year at an American university, possibly Harvard, following graduation from her Swedish high school. Conceivably she would like to talk with me during my Nobel visit. Naturally, I pledged to make myself obligingly available to explain Radcliffe's unique relation to Harvard.

The first formal event of the week was the Nobel Foundation's reception for all the year's laureates. In the grand library of the Swedish Academy, the dominant figure was John Steinbeck, who had arrived in Sweden only that morning. Though his anticipation of the honor had been keen, he was more nervous than happy, worrying about his Nobel address the next evening. William Faulkner's address of 1950 was still remembered with reverence, and Steinbeck was feeling the pressure of expectations. That evening he and his wife went off to dinner with the Swedish literary intelligentsia while I went with my fellow laureates in science to sup at the elegant naval officers' mess room on Stockholm Harbor at Skeppsholmen.

The next morning I got a sneak preview of the grandeur of the concert hall as my fellow laureates and I rehearsed the choreography of receiving a prize from the king's hands later that evening. As it usually is for most, this was to be my first experience of white-tie formality, and I was a bit self-conscious about how I looked. Betty, Dad, and I left the hotel at 3:45 P.M. to have more than enough time for me to join the backstage lineup. Precisely at 4:30 P.M. fanfare announced the arrival of the king and queen, who en-

[10] Kai Falkman went on to have distinguished diplomatic and writing careers, especially known for his biography of Dag Hammarskjöld, the Secretary General of the United Nations. Hammarskjöld, who was killed in a plane crash in 1961, was awarded a posthumous Nobel Peace Prize. (Only the Peace Prize was awarded posthumously and in 1974 the regulations were changed to bring it into conformity with the other Prizes.) Falkman is also an accomplished composer of Haiku and serves as President of the Swedish Haiku Society.

Watson undergoes an informal interview on the streets of Stockholm.

The laureates take their place on the stage. From the left, Steinbeck, Wilkins, Watson, and Crick.

Odile and Gabrielle Crick.

tered with their royal entourage and walked to their front-of-the-stage seats as the Stockholm Philharmonic Orchestra played the royal hymn. Then, with the trumpets again blaring, Max, John, Francis, Maurice, John Steinbeck, and I entered and took our seats near the front of the stage.

Before the king awarded each of the prizes, appropriate academicians read descriptions in Swedish of our respective accomplishments. To let us know what was being said, translations of their speeches had earlier been given to us. As the king handed each of us our leatherbound, individually decorated citations and gold medals, he also gave us checks in the amount of our individual shares of the prize money. From the concert hall, we went directly to Stockholm's massive 1930s city hall for the Nobel banquet, which was held in the Golden Hall. Running the entire length of the beautiful room with vaulted ceilings was a very long table where all the laureates were seated with their spouses as well as the royal entourage and members of the diplomatic corps. Placed at its center facing each other were the king and queen. I was seated on the queen's side. While Max, John, Francis, and John

Watson receives his medal, check, and certificate from King Gustaf VI Adolf.

The banquet in the Golden Hall with Crick at the center of the picture.

Steinbeck all had princesses next to them, my conversation bits were to be alternately directed to the wives of Maurice and John Steinbeck. Talking across the table made no sense, because of both its width and the alcohol-enhanced din created by more than eight hundred celebrants. During the dinner, the chairman of the Nobel Foundation, Arne Tiselius, proposed a toast to the king and queen; the king, in turn, proposed a minute of silence to honor Alfred Nobel's grand donation and philanthropy.

As soon as dessert was finished, John Steinbeck went to the grand podium overlooking the hall to deliver his Nobel address. In it he emphasized man's capacity for greatness of heart and spirit in the endless war against weakness and despair. The Cold War and the existence of nuclear weapons silently lurked behind his message of the writer confronting the human dilemma. He saw humans taking over divine prerogatives: "Having taken

John Steinbeck delivers his Nobel address. The king pays close attention.

god-like powers, we must seek in ourselves for the responsibility and the wisdom we once prayed some deity might have." Ending his oration, he paraphrased St. John the Evangelist: "In the end is the Word, and the Word is man, and the Word is with men."

I became increasingly nervous and could not listen attentively, since in just a few minutes I was to be up on the podium to offer the response of the laureates in physiology or medicine.[11] I hoped my extemporizing would rise

[11] *The first page of Watson's notes, written on letterhead from the Grand Hotel, for his Nobel address, delivered on behalf of himself, Crick, and Wilkins.*

above platitudes. Only after I was back at my seat did I relax, knowing that I had spoken from the heart. I was pleased at my last sentences, in which I had aimed for the cadence of one of JFK's better speeches. Graciously Francis then passed across the table his place card with a note on the back: "Much better than I could have done.-F." I could then enjoy John Kendrew expressing his joy at being part of a group of five men who had worked and talked together for the past fifteen years and could now come together to Stockholm on the same happy occasion. Then the party moved to the floor below for dancing, most of it done by the white ties and gowns of the Karolinska medical students.

Crick dances with Gabrielle following the banquet.

Late the next morning the laureates in science gave their formal Nobel addresses. Francis, Maurice, and I were allotted thirty minutes each. It was not an occasion for questions from our audience of mostly fellow scientists. At seven-thirty that evening, I went alone to the palace for a second royal reception at which protocol had again somehow failed to place me beside a princess.

Before lunch the next day at the American ambassador's residence, I was taken to the Wallenberg's family Enskilda Bank to exchange my advance check of 85,739 kroner for one denominated in dollars, approximately $16,500. Earlier at Nobel House, I had been given a bronze copy of my gold Nobel Medal that I could safely leave lying about my desk. There had been past thefts of the gold originals, and I was urged to keep it in a bank vault. Seemingly hundreds of photos from the past days' festivities were then shown so that I could order copies of the ones I wished. Immediately my eye alighted on one of Francis and Princess Desiree, sitting across from me at the Nobel banquet.

Watson's check for 85,739 kroners drawn on the Stockholms Enskilda Bank.

[12] James Graham Parsons was a career diplomat who had been the U.S. Ambassador to Laos and became Assistant Secretary of East Asian and Pacific Affairs. He advocated supporting the Chinese nationalist leader Chiang Kai-shek against communist expansion in Indochina but fell out of favor with the Kennedy administration. He was appointed Ambassador to Sweden in 1961.

[13] John Franklin Enders shared the 1954 Nobel Prize for Physiology or Medicine with Frederick Robbins and Thomas Weller for developing tissue culture techniques which could be used to grow polio virus, leading to the Salk polio vaccine.

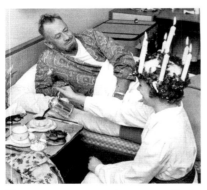

[14]*Steinbeck being served morning coffee by his St. Lucia girl. Saint Lucy was a Christian martyr dating to 283–304. Her feast day is December 13. The Swedish tradition of having a procession led by a young woman wearing a crown of candles began in the 18th century.*

Ambassador J. Graham Parsons greeted me graciously, giving no sign of the hawkish inclinations that had reputedly caused his recent banishment from the Washington corridors of Southeast Asia decision making.[12] Also welcoming us was our embassy's number two man, Thomas Enders, whom I asked if he was related to John Enders, the Harvard Medical School's polio specialist who had won the Nobel eight years earlier. In fact, this Enders was the Nobel laureate's nephew, happily no longer living behind the Iron Curtain as a junior diplomat in Poland.[13]

Nobel Week concluded traditionally on Saint Lucia's Day. Like all the laureates, I was awakened by a girl in a white robe and a crown of flaming candles, singing the Neapolitan hymn that long ago became virtually synonymous with this Swedish winter festival.[14] With our father departing that afternoon for a week in France, Betty and I again put on formal finery for the Luciaball of the Medicinska Foreningen. At dinner reindeer was served as the main course. Afterward our party moved on to a much smaller private affair that let me banter long with Ellen Huldt, a pretty dark-haired medical student, with whom I then arranged to have dinner the next night.

Before getting into a taxi to fetch Ellen, I penned a letter to President Pusey, telling him of my visit that afternoon to the Royal Palace to see Princess Christina. With my diplomatic escort, Kai Falkman, I entered one of its private reception rooms to find her with her mother, Sibylla. Over tea and cakes, I related how much I enjoyed teaching the lively students of Harvard and Radcliffe and assured the mother that her daughter would greatly enjoy a year at Radcliffe. I spent my last night in Sweden with John Steinbeck and his wife at the studio home of their friend the artist Bo Beskow. Liking *Ballet School*, one of his semi figurative blue paintings, I found its price to be within my somewhat improved means and arranged for it to be sent to Harvard. It long hung on the wall of the Biological Labs library.[15]

[15] Bo Beskow was a Swedish painter and writer. At the invitation of his friend, Dag Hammarskjöld, Beskow painted a mural in the Meditation Room of the United Nations headquarters in New York.

Appendix 1: The First Letters Describing the DNA Model

Here we reprint the first letters describing the DNA structure, one by Watson and the other by Crick.

Watson wrote *The Double Helix* 15 years after the events he describes, but he used contemporary correspondence to help him reconstruct many of the events, both scientific and not. His favorite correspondent was his sister Betty, to whom he wrote weekly when he was in Cambridge; he also wrote often to his parents back in the U.S. These letters don't describe his research to any significant extent, and while he occasionally mentions his success with TMV and his work with Bill Hayes on bacterial genetics, there is almost no reference to DNA.

In contrast, his letters to his scientific colleagues—Delbrück, Maaløe, and Luria—although less frequent, contain a great deal of science. The letter which concerns us here was to Delbrück, written on March 12, 1953. In it, Watson describes the key features of the double helix and draws a diagram of the base pairs. The letter is a technical summary of their discovery—"residues per turn in 34 Å," "stereochemical considerations," "we prefer the keto form over the enol"—written for a scientist.

The recipient of Crick's letter—the first shown here—describing the discovery was poles apart from Max Delbrück. Michael, Crick's son, was not yet 13 years old when he received a remarkable letter from his father. Dated March 19, 1953, it began "Jim Watson and I have probably made a most important discovery" and describes their structure of DNA as "…very beautiful." The letter is illustrated with sketches, although Crick "…can't draw it very well…" including a short section of double helix. The account is suitably clear and concise for a 13-year-old, and there is also a hint of fatherly sternness; Michael is admonished to "Read this carefully so that you understand it."

Crick's letter to his son Michael, March 15, 1953.

19 Portugal Place
Cambridge.
19 March '53

My Dear Michael.

Jim Watson and I have probably made a most important discovery. We have built a model for the structure of des-oxy-ribose-nucleic-acid (read it carefully) called D.N.A. for short. You may remember that the genes of the chromosomes — which carry the hereditary factors — are made up of protein and D.N.A.

Our structure is very beautiful. D.N.A. can be thought of roughly as a very long chain with flat bits sticking out. The flat bits are called the "bases". The formula is rather like this

(2)

```
            |
Sugar ——— base
            |
phosphorus
            |
sugar —— base
            |
phosphorus
            |
sugar —— base
            |
phosphorus
            |
sugar —— base
            |
```

and so on.

Now we have two ~~the~~ of these chains winding round each other — each one is a helix — and the chain, made up of sugar and phosphorus, is on the <u>outside</u>, and the bases are all on the <u>inside</u>. I can't draw it very well, but it looks

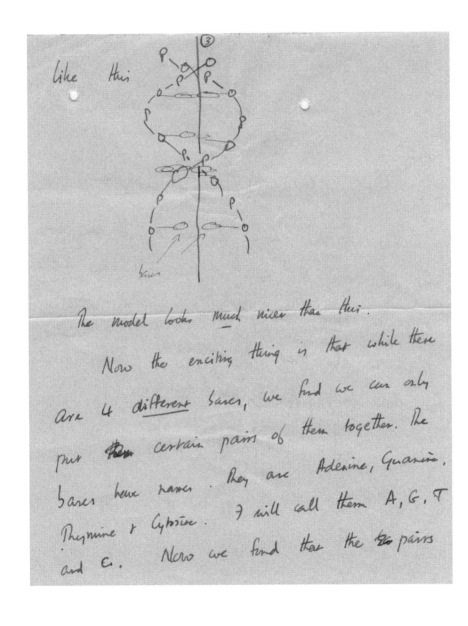

like this

The model looks _much_ nicer than this.

Now the exciting thing is that while there are 4 different bases, we find we can only put them certain pairs of them together. The bases have names. They are Adenine, Guanine, Thymine & Cytosine. I will call them A, G, T and C. Now we find that the pairs

④

we can make — which have one base from one chain joined to one base from another — are

only   A   with T

and   G   with C.

Now on one chain, as far as we can see, one can have the bases in any order, but if that order is _fixed_, then the order on the other chain is also fixed. For example, suppose the first chain goes ↓ then the second must go

A		T
T		A
C		G
A		T
G		C
T		A
T		A

⑤

It is like a code. If you ~~can~~ are given one set of letters you can write down the others.

Now we believe that the D.N.A. <u>is</u> a code. That is, the order of the bases (the letters) makes one gene different from another gene (just as one page of print is different from another).

You can now see how Nature <u>makes copies of the genes</u>. Because if the two chains unwound into two separate chains, and if each chain then makes another chain to come together on it, then because   A always goes with T, and G with C, we shall get two copies where

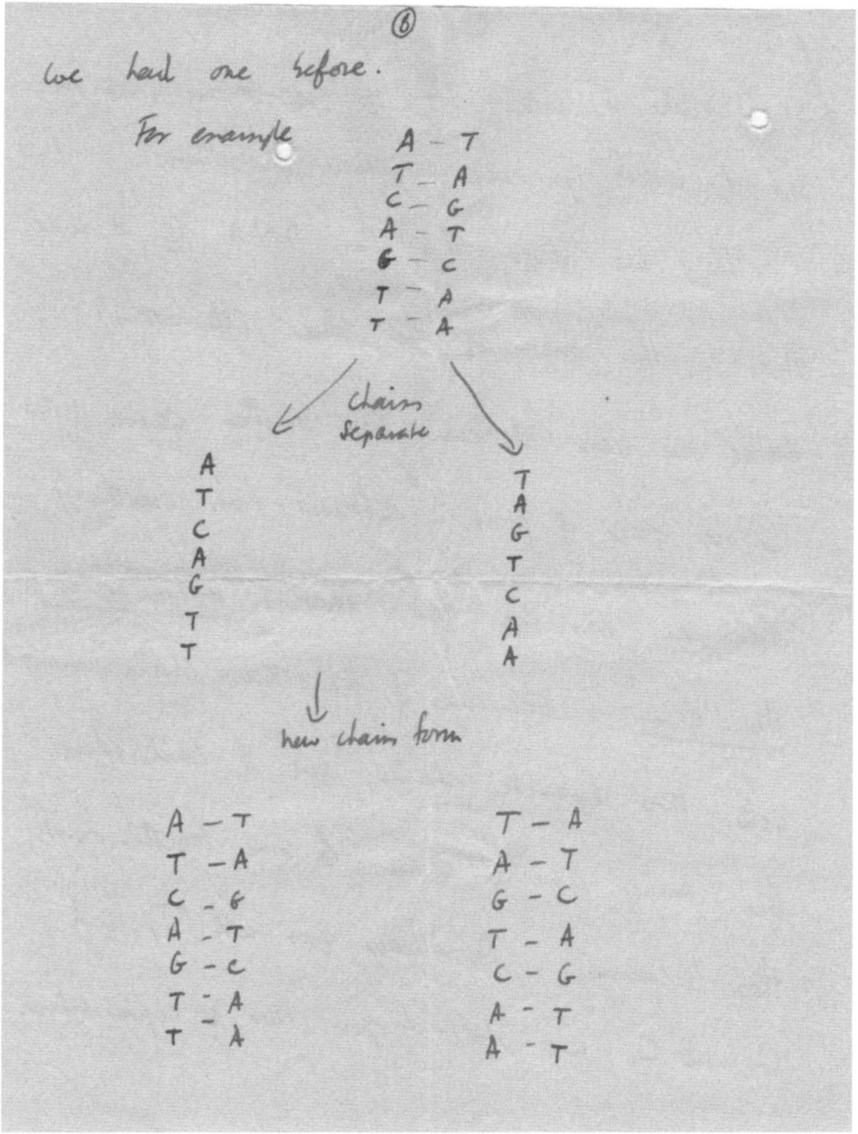

⑥

we had one before.

For example

```
A — T
T — A
C — G
A — T
G — C
T — A
T — A
```

chains
separate

```
A          T
T          A
C          G
A          T
G          C
T          T
T          C
           A
           A
```

new chains form

```
A — T          T — A
T — A          A — T
C — G          G — C
A — T          T — A
G — C          C — G
T — A          A — T
T — A          A — T
```

(7)

In other words & we think we have found the basic copying mechanism by which life come from life. The beauty of our model is that the shape of it is such that only these pairs can go together, though they could pair up in other ways if they were floating about freely. You can understand that we are very excited. We have to have a letter off to Nature in a day or so.

Read this carefully so that you understand it. When you come home we will show you the model.

Lots of love,

Daddy.

UNIVERSITY OF CAMBRIDGE   DEPARTMENT OF PHYSICS

TELEPHONE
CAMBRIDGE 55478

CAVENDISH LABORATORY
FREE SCHOOL LANE
CAMBRIDGE

March 12, 1953

Dear Max

Thank you very much for your recent letters. We were quite interested in your account of the Pauling Seminar. The day following the arrival of your letter, I received a note from Pauling, mentioning that their model had been revised, and indicating interest in our model. We shall thus have to write him in the near future as to what we are doing. Until now we preferred not to write him since we did not want to commit ourselves until we were completely sure that all of the van der Walls contacts were correct and that all aspects of our structure were stereochemically feasible. I believe now that we have made sure that our structure can be built and today we are laboriously calculating out exact atomic coordinates.

Our model (a joint project of Francis Crick and myself) bears no relationship to either the original or to the revised Pauling-Corey-Shoemaker models. It is a strange model and embodies several unusual features. However since DNA is an unusual substance we are not hesitant in being bold. The main features of the model are (i) The basic structure is helical — it consists of two intertwining helices — the core of the helix is occupied by the purine and pyrimidine bases. — The phosphate groups are on the outside (2) the helices are not identical but complementary so that if one helix contains a purine base, the other helix contains a pyrimidine. This feature is a result of our attempt to make the residues equivalent and at the same time put the purines and pyrimidine bases in the center. The pairing of the purine with pyrimidine is very exact and dictated by their desire to form hydrogen bonds. — Adenine will pair with Thymine while Guanine will always pair with Cytosine. For example

Adenine sugar

(next page)

UNIVERSITY OF CAMBRIDGE   DEPARTMENT OF PHYSICS

TELEPHONE
CAMBRIDGE 55478

CAVENDISH LABORATORY
FREE SCHOOL LANE
CAMBRIDGE

Thymine   with   Adenine

or

Cytosine   with   Guanine

While my diagram is crude, in fact these pairs form 2 very nice hydrogen bonds in which all of the angles are exactly right. This pairing is based on the effective existence of only one out of the two possible tautomeric forms - in all cases we prefer the keto form over the enol, and the amino over the imino. This is a definitely an assumption but Jerry Donohue and Bill Cochran tell us that, for all organic molecules so far examined, the keto and amino forms are present in preference to the enol and imino possibilities.

The model has been derived almost entirely from stereochemical considerations with the only x-ray consideration being the spacing between the pair of bases 3.4A which was originally found by Astbury. It turns to build itself with approximately 10 residues per turn in 34 A. The screw is right handed.

The x-ray pattern approximately agrees with the model, but since the photographs available to us are poor and negative(we have no photographs of our own and like Astbury must use Astbury's photographs) this agreement is in no way constitutes a proof of our model. We are certainly a long way from proving its correctness. To do this we must obtain collaboration from the group at Kings College London who possess very excellent photographs of a crystalline phase in addition to rather good photographs of a paracrystalline phase. Our model has been made in reference to the paracrystalline form and as yet we have no clear ideas as to how these helices can

UNIVERSITY OF CAMBRIDGE   DEPARTMENT OF PHYSICS

Telephone
Cambridge 55478

CAVENDISH LABORATORY
FREE SCHOOL LANE
CAMBRIDGE

pack together to form the crystalline phase.

In the next day or so Crick and I shall send a note to Nature proposing our structure as a possible model, at the same time emphasizing its provisional nature and the lack of proof in its favor. Even if wrong I believe it to be interesting since its provides a concrete example of a structure composed of complementary chains. If by chance, it is right then I suspect we may be making a slight dent into the manner in which DNA can reproduce itself. For these reasons (in addition to many others) I prefer this type of model over Paulings which if true would tell us next to nothing about manner of DNA reproduction.

I shall write you in a day or so about the recombination paper. Yesterday I received a very interesting note from Bill Hayes. I believe he is sending you a copy.

I have meet Alfred Tissiers recently. He seems very nice. He speaks fondly of Roscoe and I suspect has not yet become accustomed to being a Fellow of Kings.

My regards to Manny

Jim

P.S. We would prefer your not mentioning this letter to Peving. When our letter to Nature is completed we shall send him a copy. We should like to send him coordinates.

267

Appendix 2: The Lost Chapter from The Double Helix

Here we print a previously unpublished chapter from the manuscript of *The Double Helix*. It describes Watson's holiday in the Italian Alps in the summer of 1952, and would have fallen between Chapters 19 and 20 of the published book. A photo of Watson in the mountains from this trip was included in the original book (and is reproduced at the end of Chapter 19 of the current edition).

In August I stopped chasing DNA. Paula, a young Italian girl, was the most attractive object in the Italian alpine village of Chiareggio. I was there with Joe Bertani, an Italian-born phage worker back in Europe for the Royaumont gathering. Each August, Joe's family went up to a small unpretentious hotel where this year a room was reserved for me. My Italian, despite the two months in Naples the year before, was nonexistent, but at first this did not matter. Most days, Joe, his brother Alberto, and I were on the path leading onto the steep glaciers falling down from the "Disgracia," the treacherous white peak dominating Chiareggio.

For the first time since I came to Europe, I relaxed with the knowledge that I was with people who like Americans. Everywhere else the educated fraction toward whom I gravitated lived in not always silent fear that Americans were not cultivated enough to be trusted with nuclear weapons. They had to be sure that you were not an ordinary American before they felt at ease. The middle class Italian families at the hotel, however, instantly accepted me. In their eyes, Americans were generous, brave people who fought the Germans and lived in a land of unbelievable opportunity.

A local red wine, "Inferno," was drunk in healthy quantities at lunch and on the evenings of feast days, huge chocolate cream cakes filled our stomachs and prevented us from getting sick on the bubbly spumante bought by Joe's parents. The heavy food and wine consumption fitted the climbing habits of most of our hotel's guests. After breakfast, they would leisurely hike up to a nearby refuge, where during a lunch of wine, proscuitto, cheese, and fruit, they could watch roped climbers slowly come down the tricky crevasses of the Disgracia. Walking clothes did not remain on for the evening meal. Dresses and ties came out, but no one cared when I didn't change, saying that in the American mountains we were more casual.

Then already I was the local hero. Among the items for sale at a Sunday bazaar was a powerful water pistol. Quickly I used it to rid the hotel of the summer's chief nuisance, a nasty little dog owned by an unpopular guest. Before the pistol's deployment, the whining beast was a continual pest, especially after meals when we were drinking espresso. After an attack with some cheap perfume, also bought at the bazaar, the dog was no longer a problem.

Alberto and I were constantly on the lookout for Paula, who lived in the nearby chalet of her married sister. Several times each day on her way to the post or to buy the day's bread and pastries, she would parade down the single village street with two young nieces in tow. Joe made a point of not joining our gazes, since to our amazement, he did not find her simpatico. Puritanically he kept up the refrain that I was an idiot since if I could understand her Italian, I'd realize she was hopelessly bourgeois. The other hotel guests, however, did not share his sneers and seemed pleased that an Italian girl was the answer to an American's dreams.

No matter which path that Joe, Alberto, and I trudged up, I kept thinking of the moment when Paula might be my walking companion. Limited success finally came when Paula was persuaded by Alberto to join us for an afternoon excursion to a nearby refuge. The restricting prospect of being chaperoned by Joe, Alberto, and their friend Gabriella was much better than no Paula at all. Virtually all the way up and down, Paula giggled with

Paula and Gabriella, 1952.

Gabriella or Alberto, and in frustration I talked phage with Joe. The few minutes which by careful maneuvering I obtained alone with Paula were a complete failure. Long awkward silences followed Paula's inability to understand the labored phrases I had memorized out of a guide book for English tourists.

There was nothing to do but accept the fact that my mountain girlfriend would be a 16-year-old tomboy whose parents could get her into a dress only with difficulty. Mimi's two years of school English were adequate for our horsing about the hotel lobby or for racing each other up to an alp above the hotel. The last night before her family went back to Milano, we cemented our friendship by dressing for dinner and walking into the applauding dining room holding hands. Later, after an especially huge cream cake, Mimi slipped out to write me a note to go with her farewell gift to me. Raucous laughter broke out when the present, a huge pin for my long hair, was found to be the sole object in the gaudily wrapped box. The customary singing of the plaintive Italian mountain songs lasted till almost midnight, culminating in a candlelit choral procession through the village to celebrate my "betrothal" to Mimi.

Only a few days then remained before I went back down into the torrid Po valley. My thoughts, however, shifted back to England even before the mountains were out of view. The occasion was the unexpected discovery of an educated English girl in Chiareggio. Sheila's father was the Welsh Labor MP Jim Griffiths, and in her presence I could revert to the idiom of the "Economist" and the "New Statesman." We had acknowledged each other's English soon after I arrived but had not really spoken to each other till my last night in Chiareggio. That afternoon I called on the Italian family where Sheila stayed to invite her to share the spumante and chocolate cake which would mark the end of my mountain holiday. Several wine bottles were already emptied before she peered into the dining room to observe my happily inebriated companions writing a sentimental summary of my vaunted triumph over the barking little dog.

When my Italian friends finally stumbled up to bed and I was walking Sheila back to her chalet, the world which I needed to conquer fell into perspective. The folly of Bragg's and Rosy's obstructive policies must be put aside to allow the new biology of the next generation to get started. Bohr and Francis and Pauling all raced through my head as I tried to explain that I had only eight months left. Then I would be twenty-five and too old to be unusual.

Appendix 3: Watson and the Merck Fellowship Board

Watson's struggles with the Merck Fellowship Board were even more complex than his account in *The Double Helix* would suggest. He had visited the Cavendish Laboratory in early September, and made arrangements with Bragg and Perutz to take a position there in early October. He followed that visit with a letter to C. J. Lapp who oversaw the various fellowship programs administered by the National Research Council, explaining why he wanted to move to Cambridge. (We have only an undated draft of this letter, written at the beginning of October 1951.) He wrote:

> This work should be of great importance to geneticists and biochemists since, at present, progress in the biology and biochemistry of nucleic acids is seriously hindered by a lack of precise knowledge of the structure of nucleic acids…I feel that my future role as a biologist would be greatly expanded if I could have the possibility of studying in Dr. Perutz' laboratory during the coming academic year.

But changing his project and location without permission was, as Watson recognized, against the rules, and he scrupulously avoided mentioning that he had already taken steps to do so.

Kalckar was enthusiastic about Watson's plans. He wrote to the NRC on October 5, 1951, that he had encouraged Watson to use the second year of his fellowship at another laboratory and that he felt "…certain that his decision to study under the guidance of Dr. Max Perutz at Cambridge University merits full support."

Watson arrived in Cambridge on October 5 and wrote to his sister Betty on October 9 telling her that he was settling in and adjusting to the "…English way of cooking." Sometime between then and his next letter to Betty on October 16, a letter reached him from Lapp. It was not what he had

Dear Dr Lapp

I have just returned from a two and one half vacation to England and France during which time, I visited London, Cambridge, Edinburgh, and Paris. The weather was warm and sunny and provided a pleasant contrast to the wet and cold summer which Copenhagen has just experienced

I visited Cambridge in order to talk with Dr. M.F. Perutz at the Cavendish laboratory. Dr Perutz is currently studying the structure of proteins by means of x-ray diffraction patterns. In particular, he is testing the hypothesis proposed by Linus Pauling that the basic structure of most polypeptide chains is an internally hydrogen bonded helix with 3.7 amino acid residues per turn. Dr Perutz now believes that he has found a proof of the Pauling hypothesis with regard to certain globular proteins and is now engaged in testing this hypothesis upon synthetic polypeptides. I received the impression that the Cambridge group is quite optimistic about their ability to determine the precise structure of certain representative proteins and they hope in the near future to seriously attack the structure of nucleic acids. nucleoproteins.

This work should be of great importance to geneticists and biochemists since, at present, progress in the biology and biochemistry of nucleic acids is seriously limited by a lack of precise knowledge of the structure of nucleic acids. In fact, it is difficult to see how we can arrive at any satisfactory solution to the problem of the replication of macromolecules until we know their structure.

For this reason, I feel that my future role as a biologist would be greatly expanded if I could have the possibility of studying in Dr Perutz laboratory during the coming academic year. During this time, I would like to study both the theory and practice of x-ray diffraction work and if my studies are successful, attempt to study the structure of ribonucleic acid. One of the main benefits from such a year, should be a considerable increase in my facility with mathematical and physical ways of thinking.

I plan to return to Cambridge this coming week and will stay there for the remaining duration of my fellowship if the Merck Fellowship Board considers to this application in my favor. I realize that it would have been more convenient for you, if my decision could have been made earlier. I felt, however, that it would be wise to speak to Dr Perutz personally before deciding about the desirability of such a radical a change in my type of research and study. I know that I should stay in Copenhagen until I hear from the Fellowship Board

Watson's letter to Dr. Lapp, October 1951.

hoped. In the October 16 letter to Betty, Watson wrote that "…after reading my research report they could not see why I wanted to leave. I will leave the matter to Lu [Luria]. As he wanted me to work with Perutz, I know he will fight for me. I do not intend to worry about the matter."

Luria did fight for Watson. On October 18, Luria phoned Paul Weiss, the new chairman of the Fellowship Board and wrote to both Weiss and Watson on October 20. To Watson, he outlined what needed to be done (letter reproduced on page 40).

First, Luria would take the blame for persuading Watson to go to Cambridge and for not fulfilling his promise to Watson that he, Luria, would tell the Board of the move. Second, Watson was to create the impression that the Cambridge research was closely linked with what he had been doing in Copenhagen, that is doing research on viruses. He was to rush to see Kendrew and Roy Markham at the Molteno Institute and devise a scheme that would persuade Weiss that Watson wasn't really changing his research project. Markham was to be drawn into the scheme because he worked on turnip yellow mosaic virus. Markham acquiesced although he regarded these machinations "…as a perfect example of the inability of Americans to know how to behave."

Luria recognized that all this was rather underhand: "Remember that we are being somewhat unethical in making out all this a posteriori and that Englishmen are even more puritan than Americans on such matters—The problem will be not to let the Committee know that you are already in England."

Luria's October 20 letter to Weiss opens with a revealing comment on Watson: "Watson has been groomed by myself and by Dr. Max Delbrück, of the California Institute of Technology, as a boy on which we place the greatest hope for developing the work on reproduction of viruses and biological macromolecules along new and still unexplored lines."

Turning to the issue at hand, Luria wrote:

Since I am personally responsible both for encouraging Watson to make the change and for neglecting to inform the Committee of my reason for doing so,

I am writing to you now in order to suggest that the Committee, if possible, reconsider its recent action…His plans, which I suspect he failed to make adequately clear in writing to the Committee, involve both biochemical work on viruses, along lines somewhat different from those followed at Copenhagen, and the learning of the theory and techniques of X-ray diffraction analysis and their application to the virus problem…Kendrew seemed keen on the idea of a man with Watson's interests spending some time with them, and we considered the possible role of Dr. Roy Markham of the Molteno Institute, a specialist in virus nucleoprotein, in such a plan…In conclusion, I feel Watson's present plan, far from being a drift into the wilderness, is a considerate search for the type of preparation that may improve his usefulness to biology.

Weiss wrote to Watson on October 22, probably the same day as he received Luria's letter. The letter went to Copenhagen where, of course, Watson was supposed to be. It seemed that Luria's strategy was working, for Weiss wrote that he understood Watson's interest in molecular structure at Cambridge would be "merely coincidental to other work on virus nucleoproteins to be carried out at the Molteno Institute, more closely related to the line you have been following thus far." Weiss asked for details of the work Watson was to pursue at the Molteno and, more awkwardly for Watson, asked when he was intending to move from Copenhagen.

The next exchange of letters between the NRC and Watson is missing but Watson must have felt it was going well. On October 27 he wrote to Betty that "My trouble with the NRC is now almost over due to Luria's prompt and very efficient intervention. I thus believe I will be paid and so after a week's living as a poor man I can again feel mentally rich."

What Watson didn't know when he wrote so optimistically to Betty was that a letter from Weiss was winging its way across the Atlantic. Dated October 26, Watson received it on November 13 and, to judge from the reply he wrote on November 14, it cannot have been a friendly letter.

In his reply, Watson admitted that he had already moved to Cambridge as he "…felt it best to remain in the very stimulating atmosphere of Cambridge until a suitable plan for the future was agreed upon." He thought that he and Luria together had devised such a plan: "I believe Dr. Luria has indicated to you our reasons for believing that future progress in the field of virus research may benefit greatly from a synthesis of the recent advances in the field of structural protein chemistry and of current biochemical ideas concerning the structure or nucleic acids." This could be done in Cambridge because there Markham was working on viruses at the Molteno and Perutz was carrying out structural studies at the Cavendish. Watson tried to make the link explicit:

> I hope that I have made clear the connection between my proposed plan and the plan I submitted a year ago when applying for a renewal of my fellowship. The purpose of both plans is an investigation of the mechanism of virus replication. The change in emphasis from the metabolic to the structural approach is the consequence of our belief or to be more honest, our hunch, that a knowledge of the structure of nucleic acids might lead us more directly to the mechanism of replication.

Finally, as Watson puts it in *The Double Helix*, he eats crow: "I must apologise for the admittedly incoherent letter which I first wrote to you. I realize that my initial proposal was both unclear and late. I also realize that my present trip to Cambridge was premature. I can hope that the results of my future work will justify the present confusion."

More crow eating and post hoc rationalization followed in a letter of November 27 to Lapp at the NRC. Watson wrote that, having thought that Weiss agreed with his move to the Cavendish, Lapp's letter of November 21 "…objecting again to my working at Cambridge…came as a great shock to me." He had not anticipated "…that the Fellowship Board would object to my move from Copenhagen to Cambridge, and so I was surprised that you should take such a serious view of it."

Following the agreed-upon plan, he put the onus on Luria: "I did not come here on my own initiative, but on the advice of Dr. Luria."

Although Watson wrote in *The Double Helix* that telling the Fellowship Board of Kalckar's marital difficulties "…would have been not only ungentlemanly but unnecessary," it was necessary to do so now to support his case for a move to Cambridge. "…Dr. Kalckar's scientific activities were becoming severely restricted as a result of domestic difficulties, and so I did not find the encouragement and advice which I had expected…" In contrast, on his first visit to Cambridge he had found "…young and highly active research workers and the vigorous intellectual climate which was lacking in Copenhagen."

Begrudgingly he admits "…that technically I may have committed a breach of the Fellowship regulation in coming here before receiving your approval…" but justifies the transgression because what he "…did seemed so obviously in accord with the spirit of the Merck Fellowship, that I trusted you to forgive my breach of the technical regulations."

The following day, Watson brings Betty up to date, telling her that he is still in hot water, Paul Weiss having "…taken great offense…" at Watson's high-handed actions. He fears that he may lose his fellowship but with financial help from the Cavendish, he "…should be able to survive without much unhappiness."

Luria and Delbrück continued to campaign on Watson's behalf but it wasn't until January 11, 1952, that he heard from Luria that the fellowship issue was resolved. Weiss had decided that as Watson was changing project and laboratory, his application for a renewal should be treated as a new application. Watson described the consequences of this in his January 28 letter to Betty: the Merck Fellowship Board had awarded him a fellowship for eight months, thus penalizing him for four months' income.

Watson was right to think that Weiss was at the root of his troubles. On March 16, the Fellowship Board decided that the "unauthorized change made by Dr. Watson being an accomplished fact, and in sympathetic consideration of the interests of the fellow, the Board on motion made by Dr. Tatum and seconded by Dr. Clarke, confirmed unanimously the interim action taken by the Chairman [Weiss] to substitute for the original cancelled fellowship renewal a fellowship of 8 months duration to work at Cambridge University."

That was the end of the discussions with the NRC but not of Watson's interactions with Weiss, who wrote inviting Watson to a meeting in the States. As described on page 109, Watson's description of his reaction was more guarded in *The Double Helix* than in his letter to Luria:

NATIONAL RESEARCH COUNCIL
M E R C K   F E L L O W S H I P   B O A R D
Minutes of the Meeting
March 16, 1952

The meeting was called to order at 9:40 a.m., March 16, by the Chairman
of the Board, Dr. Paul A. Weiss, in the Board Room of the National Research
Council, Washington, D.C.

PRESENT: Members of the Board:

Dr. Paul A. Weiss, Chairman
Dr. Edward L. Tatum
Dr. Hans T. Clarke

Fellowship Office:

Dr. C. J. Lapp

Dr. Weiss appointed Dr. Clarke as pro tem member of the Board, since he
was past chairman of the Board from its beginning.

The following Board members were absent: Dr. Carl F. Cori, Dr. Rene J.
Dubos, Dr. John R. Johnson, and Dr. Carl F. Schmidt.

Dr. Lapp reported that there were 4 renewal applications:

Bogorad, Lawrence
Clayton, Roderick
Jagendorf, Andre
Thayer, Philip

Dr. Weiss made a report on the case of Dr. James D. Watson. Dr. Watson
left Copenhagen where he was pursuing his fellowship and went to Cambridge to
work on molecular structure analyses without knowledge or consent of the Board.
His letter of appointment specifically says that his fellowship was given to
study genetics (primarily) and biochemistry (secondarily) under Dr. Herman M.
Kalckar, his scientific adviser, in the Institute for Cytophysiology, University
of Copenhagen. In fairness to other applicants, major changes of study program
and place must be considered competitively. The unauthorized change made by Dr.
Watson being an accomplished fact, and in sympathetic consideration of the interests of the fellow, the Board on motion made by Dr. Tatum and seconded by Dr.
Clarke, confirmed unanimously the interim action taken by the Chairman to substitute for the original cancelled fellowship renewal a fellowship of 8 months
duration to work at Cambridge University.

NRC Board Minutes, March 16, 1952.

> This time in marked contrast to his [Weiss] last letter in which he told me of my immaturity. He asked me to speak at a symposium on growth in the States in late June. A very sweet letter—the bloody bastard. Naturally he assumes that I will go running home when he cut off my income and he wants to restore our friendship on his terms. His letter provided me with a wonderful opportunity to reply with sarcasm. However I wrote a very polite letter saying that I wished to continue my Cambridge work and so unfortunately could not return to the states.

Luria replied: "As for Paul Weiss, I incline to agree with your definition, although being less British than you are, I would call him a 'damn son-of-a-bitch' rather than a 'bloody bastard'."

Watson's and Kendrew's final reports on the Merck Fellowship were reviewed at the March 7, 1953 meeting of the Fellowship Board. Watson's report is headed "Cambridge Period at the Molteno Institute" and describes how "In Cambridge, I have been concerned, in collaboration with Mr. Francis Click (sic), with the structure of Tobacco Mosaic Virus." There is no mention of DNA. What a miss for the Merck Fellowship Board and the NRC! If not for Weiss's bureaucratic adherence to the regulations, they might have claimed the double helix as one of the successes of the program.

Kendrew wrote a revealing and prescient assessment of Watson's character for the NRC:

> In general we have found Watson a most stimulating colleague in our Laboratory. He has a large fund of original ideas, and great ingenuity in proposing methods to test them. He shows considerable drive in tackling a problem, and is not diverted from his object by difficulties seen ahead. There is no doubt that he is a scientist of outstanding ability.

Kendrew added a qualification:

> I would say his success is due more to a flair for ideas and significant experiment. than to the persistent and patient approach of the "plodder." His weakness lies mainly in a certain lack of system and orderliness, and of care in using physical instruments.

Which reminds us of Crick's opening line of his projected account of the discovery of the double helix: "Jim was always clumsy with his hands. One had only to see him peel an orange." Kendrew goes on:

> I repeat that we have found his general level of the highest order and we shall be fortunate to secure collaborators in the future whose standard is so high. I believe that Watson is fundamentally a man of original ideas and that as such he will go far in the scientific world, since it is those with originality who should form the leaders of scientific advance.

Appendix 4: Writing and Publishing
The Double Helix

The jacket of the first UK edition of *The Double Helix* in 1968 carries a blurb from the scientist, civil servant, and novelist C. P. Snow: "Like nothing else in literature, it gives one the feel of how creative science really happens. It opens a new world for the general non-scientific reader."

Five years earlier, Snow had written to Francis Crick urging him to write an account of the discovery of the DNA structure for a book he wanted to produce for a general audience. Crick's reply reveals enthusiasm in principle ("I think the idea of the book is an excellent one."), but he wonders how it might be done.

"[I]t could be written in two ways: either the scientists could write it, as you suggest, or it might be written by a single person who would talk to each of the scientists chosen. The ideal person for this is yourself…"

He goes on to point out the difficulty, as he saw it, of a single scientist writing the story.

> As for myself I am very much in two minds. I have given a couple of talks on how we did our DNA work. The problem in writing it up is not merely the time it would take, but also the fact that several other people (Watson, Wilkins, etc.) were involved, and everything would have to be checked and agreed with them. I can't say I am very keen.

Unbeknownst to Crick or Snow, Watson had by then already written the first chapter of what would become *The Double Helix*, including its opening line: "I have never seen Francis Crick in a modest mood." This he wrote while staying at Albert Szent-Gyorgyi's house at Woods Hole the previous summer. Having contemplated writing an account since soon after the structure was found, the impulse was finally acted upon only after giving a talk that spring at the Ambassador Hotel in New York. Collecting a prize on behalf of himself and an absent Crick, Watson gave an amusing after-dinner account of their discovery. As he later wrote of this occasion: "My unexpected candor elicited much laughter and was later praised for allowing the audience to feel like insiders in one of science's big moments." Watson felt none of the constraints voiced by Crick in his letter to Snow. Rather, he saw an opportunity to write the story in the style Truman Capote would later characterize as the "nonfiction novel."

A variety of commitments, including those associated with winning the Nobel Prize that autumn, kept Watson from writing more than the opening chapter for almost a year. And even then, in the summer of 1963, he wrote only a couple more chapters because he'd also embarked on what would become his influential textbook *Molecular Biology of the Gene* (published in 1965); he largely set *The Double Helix* aside until the textbook was done. The bulk of the remaining writing was completed while he was on sabbatical from Harvard, at Cambridge University in the summer of 1965, staying in rooms arranged for him by Sydney Brenner in King's College, and finally that Christmas at Carradale, the Mitchison's Scottish house that also features in the story. Being back in Cambridge allowed him to confirm aspects of the story with Crick, whose secretary typed up the chapters as Watson completed them. At this stage Crick seemed at worst ambivalent about the project.

Watson initially entitled his book *Honest Jim*. This choice originated in Willy Seeds' remark to Watson as they passed each other in the Alps in 1955, an encounter described in the Prologue. But the title was also a nod to Kingsley Amis' *Lucky Jim* and Joseph Conrad's *Lord Jim* whose themes Watson saw mirrored in his book—from the faintly comic aspects of immediate postwar English academic life portrayed by Amis to issues of character investigated by Conrad. The link to *Lucky Jim* has stuck, despite the final title of the book changing to *The Double Helix*. Analogies with Amis' book have even extended to perceived similarities in how Amis portrayed the poet Philip Larkin's long-term lover Monica Jones as the character Margaret Peel, and how Watson portrayed Rosalind Franklin.

The first publisher to become aware of the book was Houghton Mifflin who got an early look at a partial manuscript through a social contact. But a meeting with the publisher's lawyers revealed that they were leery of the lawsuits they feared the portrayal of Crick and others might justify. Unimpressed, Watson returned to Harvard "suspecting that Houghton Mifflin's risk aversion could not allow them beyond the jeopardy that might attend issuing further editions of Roger Tory Peterson's bird guides." The next publisher was more promising, at least initially. This was Harvard's own University Press (HUP), whose Director Tom Wilson was immediately—and remained—a champion of the book.

Visiting London soon after learning HUP would like to take the book, Watson lunched at Wheeler's famous fish restaurant in Dover Street with Peter Pauling and some young scientists, including the crystallographer Tony North. Pauling and the others were all based at the Royal Institution, an establishment at that time presided over by Sir Lawrence Bragg who had moved there after leaving the Cavendish in 1954. Over a few bottles of wine, Watson showed them his manuscript and admitted that he was worried how Bragg might respond to it. It was North who came up with the idea of asking Bragg to write a foreword, a clever plan in conception, though nerve-wracking to carry out. Indeed, it was not until a couple of months later that Watson found the opportunity and courage to broach the matter with Bragg.

By then Watson was staying in Geneva, from where he took two trips to London to see Bragg—the first to give him the manuscript of *Honest Jim*, and the second to learn whether he would be willing to provide a foreword. Bragg's initial fury on reading how Watson had presented him and others was apparently cooled by talking it through with his wife Alice, and he came to appreciate that what Watson had produced was novel and worthwhile. He agreed to provide the foreword, even while acknowledging jokingly—ominously—that this action would of course deny him the opportunity to sue for libel.

The significance of getting Bragg's foreword was immense, and not lost on Watson. Had he not gotten it, the chances are high that the book would never have been published, especially had Bragg instead joined forces with Crick and Wilkins who would come to object strongly to publication of the book, as we shall see.

It was at this stage that Watson sent a complete draft to Crick for comments. For now Crick's complaints were largely restricted to points of fact, though even in his first response he makes clear that "This should not be taken to imply that I agree with the remainder of the manuscript—there are quite a number of judgments which I believe to be false which are not strictly matters of fact," March 31, 1966.

Nevertheless, his concern seemed to be with improving the accuracy of the manuscript rather than objecting on principle to its publication. He was also not happy with the title. He felt *Honest Jim* gave the smug impression that Watson alone among them was offering up the honest truth. But when an alternative title was produced—*Base Pairs*—his objections were even stronger (September 27, 1966): "I am sure you will see that everyone will identify the two of us as at least one of the pairs, and I do not see why I should have a book published in which I am described as 'base'." Indeed, faced with this alternative, his opinion of the original title softened considerably, if slightly sarcastically: "Personally, I thought the title 'Honest Jim' an excellent one; I cannot see any reason why you would not wish to be called 'honest'...You had better think of another [title], or go back to 'Honest Jim'." The publishers were also squeamish of these titles and in the end Watson accepted their suggestion of the less colorful *The Double Helix*.

The nature of Crick's criticisms changed sharply just a week after his letter about the title. The focus was no longer on modifying the manuscript. Now he was against publication altogether. The change in attitude was probably fueled by a visit to Maurice Wilkins who had by then seen a draft as well. Wilkins was against publication and presumably discovering a shared indignation emboldened each of them to take a stronger line with Watson. Thus, on October 3, 1966, Crick again wrote to Watson:

> I have now looked again at your book, and being somewhat less harassed than when you sent me the earlier version in the Spring I have had time to reflect on the whole matter. I have also discussed it with Maurice. Reluctantly I have come to the conclusion that I cannot agree to its publication. I have two reasons for this. The first I have already told you in outline. There is far too much gossip and the intellectual content is too low. The second reason is that, as you

know, I have very largely avoided personal publicity in the last few years. If I agreed to the publication of your book I could no longer do this.

The letter continues to articulate his anxieties and disapproval, ending:

Finally I should point out to you that your book, far from benefiting science, may actually do it harm by setting a most dangerous precedent. People will think twice about working together if highly personal accounts of their collaboration are liable to be published. I think the unwritten convention that discourages scientists from doing this is a wise one.

I do very much regret not having taken a firm line earlier. I have always made plain to you my dislike of the whole idea of your book, and for this reason refused to read your earlier drafts. The manuscript you sent me this Spring arrived at a most unfortunate time when Odile was seriously ill. I could not consult Maurice as you had not at that time shown the manuscript to him. I have now discussed the whole matter with him and find that he agrees with me that your book should not be published.

I have therefore written to this effect to the Harvard University Press, (Copy enclosed), and have sent them also a copy of this letter. I do hope under the circumstances that you will have the good sense not to proceed with publication, although I realize this will be a disappointment for you.

A few days later, a letter arrived from Wilkins, who, despite his characteristically torn tone, also came down decisively against publication (October 6, 1966). (The letter is reproduced on the following pages.)

Dear Jim,

When I first heard you were writing "Honest Jim" I was very doubtful about the desirability of its being published (my letter of February 18th). I was, however, very interested to read it and to see to what extent modification might improve it. Now, faced with the semi-final draft and the publishers' forms for signature, I have thought the whole matter over again and find myself taking the views I expressed in the beginning. To suggest that a book should be suppressed is something one does not like to do but I am oppressed by thoughts of the undesirable effects of publishing the book. It is, in my opinion, unfair to me, and this has made it more difficult to sort out my thoughts.

Towards the end of the letter he reveals his dread of publicity, and also refers to Franklin's position in the matter.

MEDICAL RESEARCH COUNCIL

BIOPHYSICS RESEARCH UNIT

Telephone :
TEMple Bar 8851

DEPARTMENT OF BIOPHYSICS,
KING'S COLLEGE,
26-29 DRURY LANE,
LONDON, W.C.2.

6th October 1966

Professor J.D.Watson,
Harvard University,
The Biological Laboratories,
16, Divinity Avenue,
Cambridge 38,
Mass., U.S.A.

Dear Jim,

When I first heard you were writing 'Honest Jim' I was
very doubtful about the desirability of its being published (my
letter of February 18th).   I was, however, very interested to
read it and to see to what extent modification might improve it.
Now, faced with the semi-final draft and the publishers' form
for signature, I have thought the whole matter over again and
find myself taking the views I expressed in the beginning.   To
suggest that a book should be suppressed is something one does
not like to do but I am oppressed by thoughts of the undesirable
effects of publishing the book.   It is, in my opinion, unfair to
me, and this has made it more difficult to sort out my thoughts.

I am with you in being tired of polite covering-up and
misleading inadequate pictures of how scientific research is
done, but I think there is sense in the way scientific people —
and academics generally — have tried to shield each other from
vulgar gaze.   With increasing interest in science there is going
to be more and more pressure to take the lid off, but if the old
conventions are to be replaced it is important to choose carefully
how.   There is already much spilling of beans in military memoirs
and by lawyers, politicians and journalists, and confidential
matters are revealed increasingly soon after important events.
Some tendency this way is probably inevitable in the academic
world but do we want to accelerate it?   Because you are a
scientist of the very highest standing, a book from you would
be a sign to others to go ahead with accounts of their feelings
and impressions concerning their work and collaborations.
Meanwhile, scientific research becomes of increasing social
importance and as a human activity badly needs scientific study —

Wilkins' letter to Watson.

in particular, the history of contemporary science needs developing. Clearly this needs to be done in a scholarly way. I think publication of your book would impede such development.

The book would present to non-scientists a distorted and unfavourable image of scientists. The DNA story is not typical of scientific discovery; for one thing it was unusually involved with personal difficulties. Most top scientists are fairly civilised, but your book, though you may not intend it, would give many people an impression of Francis as a feather-brained hyperthyroid, me an overgentlemanly mug and you an immature exhibitionist! This would not be fair to any of us or to scientists in general. I think you will agree that the barrier between arts and science is a bad thing and that there is real need to establish, in the intellectual and academic world, science as a cultural activity deserving respect. Most people realise that scientists have human failings like everyone else, and that scandal and intrigue is often present in their world, but I think your book overemphasises this. It would be undesirable too if you gave the impression you _enjoyed_ revealing scandal.

The book is likely to arouse considerable interest and cause newspaper people, etc., to pester me to confirm or deny what you say. I do not want to be pestered and I do not want to be forced into a position where I might say that you were an eccentric who should not be taken seriously. Nor do I want to stand on one side while Rosalind is discredited. She was my colleague and, however just your account of her might be, I cannot approve its publication: she would certainly not if she were alive.

None of my objections applies to a thorough study of the whole history. If writing your book stimulates such study it will have been very worthwhile.

Yours *Maurice*

M.H.F.Wilkins

P.S. I enclose copy of my letter to H.O.P.

Wilkins' letter to Watson.

The book is likely to arouse considerable interest and cause newspaper people, etc., to pester me to confirm or deny what you say. I do not want to be pestered and I do not want to be forced into a position where I might say that you were an eccentric who should not be taken seriously. Nor do I want to stand on one side while Rosalind is discredited. She was my colleague and, however just your account of her might be, I cannot approve its publication: she would certainly not if she were alive.

Watson replied to Crick on October 19: "I am naturally disappointed in your letter…" In response to Crick's complaints about there being more gossip than science Watson writes: "Your argument that my book contains far too much gossip and not enough intellectual comments misses entirely what I have tried to do. I never intended to produce a technical volume aimed only at historians of science. Instead I have always felt that the story of how the interactions of me, you, Maurice, Rosalind, Bragg, Linus Pauling, Peter P., etc., finally knitted into the double helix was a very good story that the public would enjoy knowing…Someday, perhaps you or Maurice, but if not some graduate student in search of a Ph.D. will write a balanced scholarly historical work…" Watson also tries to impress Crick with the number of people—fifty or so—who had enjoyed the manuscript and believed it should be published. He ends by urging Crick to allow publication soon.

From this point the disagreement rapidly escalated. Crick's fury mounting, he wrote first to Tom Wilson, head of HUP, and finding no satisfaction on that front, to Nathan Pusey, President of Harvard University itself, demanding that he intervene to stop publication.

Watson had been making changes to the manuscript in response to comments he received, including almost all of Crick's earlier specific criticisms—errors of fact were corrected (we noted some examples in annotations to the text, for example see page 59), and some disagreeable phrases were removed and perceived misrepresentations recast. In addition, in response to an insightful suggestion by his editor at HUP, Joyce Lebowitz, Watson added an Epilogue, mainly about Rosalind Franklin. In this Watson reiterates that the book intentionally records events and personalities as they appeared to him at the time, as a 23-year-old American in Cambridge in the early 1950s. He acknowledges that this resulted in rather a distorted view of Rosalind Franklin, and as the only character not still alive, he takes the opportunity to correct some misrepresentations and to celebrate the important work on TMV that she later carried out before her untimely death.

But in the face of Crick's blanket refusal to sanction publication, irrespective of the number of specific corrections he made, Watson wrote a short note to him on November 23, 1966:

Dear Francis:

I am troubled very much by your hostile reactions to "Honest Jim." The thought of our long, most productive, and thoroughly enjoyable friendship coming to an unnecessary end thoroughly depresses me. But you offer no possibility for compromise and tell me that the book is not only a stab-in-the-back invasion of your privacy but in bad taste and poorly written. But, as I think it is a good book and in no way harms you or your reputation, I cannot bring myself to accept your request. I say this with much regret for on most occasions I have found your judgments sensible and to the point.

But at this moment I regretfully cannot follow your counsel.

Yours sincerely

J. D. Watson

A line was drawn here. Watson determined to publish even without Crick's approval. But things were far from over.

More letters flew back and forth across the Atlantic. And other members of the scientific community got involved, mostly in support of publication. Some—like J. D. Bernal, George Klein, Richard Feynman, and John Maddox—agreed the book was something extraordinary, un-put-downable, and despite some qualms and anxieties (and concerns for Watson's well-being should the book appear) all supported publication. Linus Pauling (shown a copy by his son Peter) was not happy, but never tried seriously to block publication. A couple of eminent colleagues of Watson's at Harvard—Paul Doty and John Edsall—were pulled into the dispute, partly in response to Crick's letters to President Pusey. Both wrote letters to Crick which made clear that, while having some sympathy for his position, on balance they felt publication was justified.

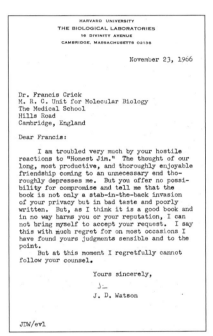

Watson to Crick.

BIRKBECK COLLEGE
(UNIVERSITY OF LONDON)
MALET STREET
W.C.1
LANGHAM 6622

DEPARTMENT OF CRYSTALLOGRAPHY
PROFESSOR J. D. BERNAL, M.A., F.R.S.

Dr. J. C. Kendrew, F.R.S.,
M.R.C. Laboratory of Molecular Biology,
Hills Road,
CAMBRIDGE.   20th December, 1966.

Dear Kendrew,

    I have now read the book entitled <u>Base Pairs</u> by Watson.
It is an astonishing production, I could not put it down.
Considered as a novel of the history of science, as it should
be written, it is unequalled. It is as exciting as Martin
Arrowsmith but has the advantage of being about the history
of a real and very important discovery. It raises many vital
problems, not only about the structure of DNA but about the
mechanism of scientific discovery which he shows up in a very
bad light. I am astonished that it is allowed to be published.
In England it would be libellous in many places, but I imagine
U.S. laws are different.

    As someone who comes into it by implication but not directly -
I never met Watson before the discovery but if I had I could have
told him quite a lot - what impressed me most is that he did not
know, and apparently never tried to find out, what had been done
already in the subject. He is particularly unfair on the
contribution of Rosalind Franklin and does not mention her
projection of the helical DNA structure showing the external
position of the phosphate groups. I need not mention the complete
absence of a reference to the work of Furberg which contains all
the answers to the structure except one vital one - the double
character of the chain and the hydrogen bond base pair linkage.
Effectively, all the essentials of the structure were present in
Astbury's original studies, including the negative birefringence
and the 3.4 Å piling of the base groups. I should add in my own
defence that my weakness was in what he calls the English habit
of respect for other peoples work. There was a tacit understanding.
I dealt with biological crystalline substances and Astbury dealt
with messy substances. Nucleic acids came clearly in the second
category. It was not that I considered them unimportant but it
was not my responsibility. I was certainly wrong in this. Astbury
was quite clearly incapable of working out the structure. The
genetic importance of DNA was apparent to me long before from the
work of Caspersson which, Watson hardly mentions. Watson and
Crick did a magnificent job but in the process were forced to
make enormous mistakes which they had the skill to correct in time.
The whole thing is a disgraceful exposure of the stupidity of
great scientific discoveries. My verdict would be the lines of
Hilaire Belloc'

         "And is it true? It is not true!
         And if it was it wouldn't do."

Bernal's letter to Kendrew commenting on the manuscript Base Pairs.

Dr. J. C. Kendrew, F.R.S.   20th December, 1966.

I am sure this publication of <u>Base Pairs</u> will cause a lot of heart-burnings in scientific circles and particularly in England but it makes very good reading and I think it would make an even better film because it is so alive and dramatic.

I will keep it for another few days and then will send it back to you. There is a page missing and another illegibly copied.

I enjoyed our conversation the other day very much.

Yours sincerely,

J. D. Bernal

J. D. Bernal.

Bernal to Kendrew, page 2.

Doty's letter of March 16, 1967 points out that part of the problem stemmed from Crick having not objected to publication until very late in the process. Earlier, as Doty pointed out, it had appeared that Crick "had no serious objection, displaying as you did then a rather detached, patient and bemused attitude." Doty worried that now the manuscript had been widely disseminated (and almost everyone who read it had agreed it should be published), suppressing it would now be awkward to say the least.

Edsall had objected to Crick's claim that surely Watson could not publish without his (Crick's) approval. Edsall agreed that a collaborator's approval was essential if the publication were a scientific paper, but surely not a memoir, pointing out that such a rule did not apply in any other field.

Watson was nevertheless worried that the pressures were building to a point that Harvard would be unable to go ahead. He also worried that Bragg might regret having written the foreword, and offered him the chance to re-

from JOHN C. KENDREW, *The Guildhall,* 4 *Church Lane, Linton, Cambridgeshire*

23.4.67   *Linton* 545

Congratulations on becoming an Hon Fellow & hope we shall see you here soon to enjoy it.

Francis showed me his last salvo to you about Ernest J. — my impression is that he is now giving up the struggle.   I had the Braggs staying & helped him re-draft the introduction   Hope all now straightforward   In haste
J

Kendrew's card to Watson about helping Bragg rewrite the introduction.

move it. Bragg was indeed feeling rather unhappy about the unfolding conflict over the book, but he said he would still provide the foreword (if in slightly modified form) as long as Watson made further specific changes to the text of the book. As in earlier cases, Watson agreed to these requests for particular defined changes. The Braggs stayed with John Kendrew in April 1967, and Kendrew helped Bragg redraft the foreword.

Crick's last blast came in a letter of April 13, 1967 (see extracts from this letter on pages 296–298). Addressed to Watson, it was copied to ten others by then involved in the dispute: Pusey, Bragg, Wilkins, Pauling, Wilson, Edsall, Doty, Kendrew, Perutz, and Aaron Klug. The six-page letter reiterates all his objections to the book both as history and as autobiography. He again points to important scientific details that are left out, and the many peripheral incidents that are included but which Crick feels are irrelevant (for example, Watson's Christmas at Carradale). He rails against what he sees as "the history of scientific discovery [being] displayed as gossip":

> Anything with any intellectual content, including matters which were of central importance to us at the time, is skipped over or omitted. Your view of history is that found in the lower class of women's magazines.

Crick also objects to the book as autobiography on grounds of accuracy and taste. He additionally takes on, and dismisses, the arguments raised by others in defense of the book—including those from Doty and Edsall. He even tries to dissuade Watson from proceeding with publication by claiming that in doing so he will humiliate himself:

> I do not think you realize what others will see in it. One psychiatrist who saw your collection of pictures said it could only have been made by a man who hated women. In a similar way another psychiatrist, who read "Honest Jim" said that what emerged most strongly was your love for your sister. This was much discussed by your friends while you were working in Cambridge, but so far they have refrained from writing about it. I doubt if others will show this restraint.

Things having reached such a crescendo, Harvard decided it couldn't publish a book that was causing so much division between scientific colleagues. Tom Wilson, the director of the Press, had already planned to step down from HUP at this time, and did so, joining a newly established commercial publisher called Atheneum. With Watson's and Harvard's blessing, he took *Honest Jim* with him.

But lawyers at Atheneum also had concerns over what they saw as the libelous tone of the book, and hoped Watson might be persuaded to sanitize the text further. The proposals were maddeningly technical—for example, changing "I have never seen Francis Crick in a modest mood" to the legally more defensible "I can't ever remember seeing Francis Crick in a modest mood." In the end Watson hired the renowned freedom-of-speech lawyer Ephraim London who was able to reassure Atheneum that *The Double Helix*—as it had now become—was not libelous of anybody.

The book was serialized in *The Atlantic Monthly* magazine in their January and February issues of 1968, just before the book appeared in the U.S.

Here we reproduce three of the six pages that constituted Crick's final attempt to persuade Watson not to publish *The Double Helix*.

MEDICAL RESEARCH COUNCIL

Telephone :
Cambridge 48011

LABORATORY OF MOLECULAR BIOLOGY,
UNIVERSITY POSTGRADUATE MEDICAL SCHOOL,
HILLS ROAD,
CAMBRIDGE.

13th April 1967.

Dr J.D. Watson,
Harvard University Biological
    Laboratories,
16 Divinity Avenue,
Cambridge, Mass. 02139,
U.S.A.

Dear Jim,

    The new version of Honest Jim is naturally a little better, but my basic objections to it remain the same as before.   They are:

I. The book is not a history of the discovery of DNA, as you claim in
       the preface.   Instead it is a fragment of your autobiography
       which covers the period when you worked on DNA.

    I do not see how anybody can seriously dispute this, for the following
       reasons:-

    a) Important scientific considerations, <u>which concerned you at the
          time</u>, are omitted.   For example the work of Furberg, which
          established the relative configuration of the sugar and the
          base.   There are many other examples.
    b) Such scientific details that are mentioned are referred to rather
          than described.   For example, you do not explain exactly why
          you got the water content of DNA wrong, nor make it clear that
          if there had been so little water electrostatic forces were
          bound to predominate.   You do not mention that Pauling worked
          from an old X-ray picture of Astbury's which had both the A
          and B pictures on the same photograph.   There are many other
          examples.
    c) The thread of the argument is often lost beneath the mass of
          personal details.   For example I asked both Bragg and Doty
          the following question.   "Since we had realized that 1:1

Crick's final salvo, page 1.

- 2 -

<u>Dr J.D. Watson.</u>   <u>13th April 1967.</u>

    base ratios mean that the bases went together in pairs why
did we not immediately use this idea when we started model
building the second time?"   Neither could give the correct
answer.
d) No attempt is made to ask or answer questions which would
    interest the historian (such as the one above). For example,
the advantages or disadvantages of collaboration, or when the
structure would have been solved if we had not solved it.
Nothing is said about the importance of the MRC, nor why
they decided to finance "biophysics" after the war.
e) Gossip is preferred to scientific considerations.   For example,
    you explain how Bragg and I had a misunderstanding but you
omit to say what the scientific issue was.
f) Much of the gossip and even some of the science is irrelevant to
    a history of DNA.   For example, your work on TMV and bacterial
genetics is only of marginal importance to the main theme.
Whole chapters, such as Chapter 15 on your visit to Carradale,
are irrelevant as far as DNA is concerned.   Even when
personal matters should be mentioned they are described in
quite unnecessary detail.
g) Absolutely no attempt is made to document your assertions, many
    of which are not completely accurate because of your faulty
memory.   You have not troubled to consult documents which
you could easily lay your hands on, nor have you made available
to others the documents you yourself have, such as the letters
you wrote at the time to your mother, which are in fact not
even mentioned in the book.   Dates are given in the book only
very casually.

    It is thus absolutely clear that your book is not history as normally
understood.   However once it is realized that it is not history
but a part of your autobiography many of the points made above
become irrelevant.   Unfortunately you yourself claim it as
history, and the misguided but worthy people who are supporting
you in publishing it also use this as their major excuse for
publication.

    Should you persist in regarding your book as history I should add
that it shows such a naive and egotistical view of the subject
as to be scarcely credible.   Anything which concerns you and
your reactions, apparently, is historically relevant, and anything else is thought not to matter.   In particular the history

Crick to Watson, page 2.

```
                              - 6 -


Dr J.D. Watson.                                13th April 1967.


      There is no reason why your book, as it stands, should not be
made available to selected scholars, provided any documents you may
have (such as your letters to your mother) which bear on the subject
are also made available at the same time.

      My objection, in short, is to the widespread dissemination of a
book which grossly invades my privacy, and I have yet to hear an
argument which adequately excuses such a violation of friendship.   If
you publish your book now, in the teeth of my opposition, history will
condemn you, for the reasons set out in this letter.

      I have written separately to Wilson pointing out several cases
of factual errors in your latest draft.   I enclose a copy of my letter
to him.

                              Yours sincerely,


                                  Francis


                              F.H.C. Crick.


Copies to: President Pusey.
           Sir Lawrence Bragg.
           M.H.F. Wilkins.
           L. Pauling.
           T.J. Wilson.
           J.T. Edsall.
           P. Doty.
           J.C. Kendrew.
           M.F. Perutz.
           A. Klug.
```

Crick to Watson, page 6.

There was a slight delay in publishing the UK edition (by Weidenfield and Nicolson) partly due to Watson's demand that the original dust jacket be destroyed and replaced because of comments on the back cover copy about Crick ("Which winner of the Nobel Prize has a voice so loud that it can actually produce a buzzing in the ears?" etc.).

The book was dedicated to Naomi Mitchison, to her great delight (she would later return the favor, dedicating her science fiction novel *Solution Three* "To Jim Watson, who first suggested this horrid idea."). Upon seeing the published copy of *The Double Helix*, she wrote to Watson:

My dear Jim

It's a great thrill seeing Honest Jim (though I think the present is the more definitely romantic title – it sounds just like a Celtic fairy tale) actually in print. I've seen quite a lot of reviews including that smashing one in New Scientist and I feel almost as if it was my child.

It has had one curious effect, that it has detached my grandson Graeme, who read the first draft, from topology and has sent him along to Brenner. But really it is a break, something nobody has done before. Perhaps you will never write this sort of thing again, that doesn't matter, you've done it once for all.

I do hope you and Francis are not fighting. It seems a bit silly. There are so many real things to fight about and Francis comes out pretty well, considering.

Carradale has adequate heating these days…and an under-carpet hot blanket thing, so you would only need to lie on it and all would be well…

Much love and thanks – it really is flattering to have my name on it!

Naomi

The book was reviewed widely, in both the popular and the scientific press, by such illustrious names as Peter Medawar and Jacob Bronowski. Its reception was largely but not uniformly positive, and is discussed further in Appendix 5. Many of the more prominent reviews are reprinted in the Norton Critical Edition of *The Double Helix*.

One review, perhaps inevitably negative, was by Erwin Chargaff in *Science* magazine. Chargaff denied permission to have his review reprinted in the Norton edition, but it is reprinted here in the next Appendix. In his review, Chargaff claimed that the MRC report shown to Watson and Crick by Perutz was confidential, and thus Perutz had committed a grave error. This accusation triggered letters to *Science* from Perutz, Wilkins, and Watson, and these too are reprinted in Appendix 5.

And what of Crick's reaction, once the book finally appeared? He soon let go of his previously passionate objections, and his friendship with Watson survived the ordeal. Watson and his new wife Liz stayed with the Cricks in Cambridge in the summer of 1969, and in 1972 Crick agreed to appear in a BBC documentary about the discovery of the structure of DNA. The two of them were shown in their old Cambridge haunts, including The Eagle pub, and were filmed retelling the story of how their DNA work had unfolded. Crick even refers to Watson's book a number of times. And two years later, in an article in *Nature* written as part of the celebrations surrounding the 21st birthday of the double helix, Crick pokes gentle fun at the style of *The Double Helix*: "As to a book I confess I did get as far as composing a title (*The Loose Screw*) and what I hoped was a catchy opening ('Jim was always clumsy with his hands. One had only to see him peel an orange.') but I found I had no stomach to go on."

Later Crick wrote his own memoir of his life in research, *What Mad Pursuit*. In the chapter on books and films about the discovery of DNA structure, he comments on *The Double Helix*:

> I recall that when Jim was writing the book he read a chapter to me while we were dining together at a small restaurant near Harvard Square. I found it difficult to take his account seriously. "Who," I asked myself, "could possibly want to read stuff like this?" Little did I know! My years of concentration on the fascinating problems of molecular biology had, in some respects, led me to live in an ivory tower. Since all the people I met were mainly concerned with the intellectual interest of these problems, I must have tacitly assumed that everyone was like that. Now I know better. The average adult can usually enjoy something only if it relates to what he knows already, and what he knows about science is in many cases pitifully inadequate. What almost everybody is familiar with is the vagaries of human behavior. People find it much easier to appreciate stories of competition, frustration, and animosity, against a background of parties, foreign girls and punting on the river, than the details of the science involved.

> I now appreciate how skillful Jim was, not only in making the book read like a detective story (several people have told me they were unable to put it down) but also by managing to include a surprisingly large amount of the science, although naturally the more mathematical parts had to be left out.

The book was an immediate best-seller, and, while never reaching the top, it remained on *The New York Times* best-seller list for 16 weeks, and went on to sell over a million copies and be translated into more than 20 languages. Its significance was also recognized when it placed seventh on the Modern Library list of the 20th century's best works of nonfiction and in 2012 was listed among the 88 books selected by the Library of Congress as "Books that Shaped America."

Appendix 5: Chargaff's Review and the Ensuing Controversy

The controversy surrounding the publication of *The Double Helix*, recounted in the previous appendix, served to ensure that the book was one of the publishing sensations of 1968. There were many reviews, in publications as diverse as *Nature* and *The Daily Mail* in England, and *Science* and the *Chicago Sun Times* in the U.S.

The opinions of the reviewers, as discussed by Gunther Stent (1980), were equally diverse. Many commented on the picture Watson painted of the way research is done, some scandalized by his portrayal, others celebrating it. The book's literary quality was assessed, some condemning it as trivial, others saying that Watson deserved a Nobel Prize for Literature. There were pronouncements on whether the book was destined for the best-sellers' list or the remainders table.

Stent could not reprint Erwin Chargaff's review in *Science*, which is on the following pages, together with the letters from Perutz, Wilkins, and Watson in response to it. As will be guessed from Chargaff's appearances in *The Double Helix*, he was not enthusiastic about the book.

Chargaff used his review to excoriate contemporary scientists whose lust for fame was distorting the nobility of the research endeavor. The heroes of Watson's book, "…a new kind of scientist, and one that could hardly have been thought of before science became a mass occupation, subject to, and forming part of, all the vulgarities of the communications media." He wrote that he knew of "…no other document in which the degradation of present-day science to a spectator sport is so clearly brought out."

Chargaff was not impressed by the style of the book. It lacked "…the champagne sparkle of Sterne's garrulous prose," and bubbles of soda water were the best Watson could achieve. Indeed, Chargaff thought the style was that of a "molecular Cholly Knickerbocker," the pseudonym of the syndicated gossip columnist of the Hearst newspapers.

Book Reviews

A Quick Climb Up Mount Olympus

The Double Helix. A Personal Account of the Discovery of the Structure of DNA. JAMES D. WATSON. Atheneum, New York, 1968. xvi + 238 pp., illus. $5.95.

Unfortunately, I hear it very often said of a scientist, "He's got charisma." What is meant by "charisma" is not easy to say. It seems to refer to some sort of ambrosial body odor: an emanation that can be recognized most easily by the fact that "charismatic" individuals expect to be paid at least two-ninths more than the rest, unless Schweitzer or Einstein chairs are available. But what does one do if two men share one charisma?

This would certainly seem to be the case with the two who popularized base-pairing in DNA and conceived the celebrated structural model that has become the emblem of a new science, molecular biology. This model furnishes the title of this "personal account," and Watson describes it, without undue modesty, as "perhaps the most famous event in biology since Darwin's book." Whether Gregor Mendel's ghost concurred in this rodomontade is not stated. The book as a whole testifies, however, to a regrettable degree of strand separation which one would not have thought possible between heavenly twins; for what is Castor without Pollux?

This is the beginning of chapter 1 of Watson's book:

I have never seen Francis Crick in a modest mood. Perhaps in other company he is that way, but I have never had reason so to judge him. It has nothing to do with his present fame. Already he is much talked about, usually with reverence, and someday he may be considered in the category of Rutherford or Bohr. But this was not true when, in the fall of 1951, I came to the Cavendish Laboratory of Cambridge University. . . .

As we read on, the impression grows that we are being taken on a sentimental journey; and if the book lacks the champagne sparkle of Sterne's garrulous prose, it bubbles at least like soda water: a beverage that some people are reported to like more than others. The patter is maintained throughout, and habitual readers of gossip columns will like the book immensely: it is a sort of molecular Cholly Knickerbocker. They will be happy to hear all about the marital difficulties of one distinguished scientist (p. 26), the kissing habits of another (p. 66), or the stomach troubles of a third (p. 136). The names are preserved for posterity; only I have omitted them here. Do you wish to accompany the founders of a new science as they run after the "Cambridge popsies"? Or do you want to share with them an important truth? "An important truth was slowly entering my head: a scientist's life might be interesting socially as well as intellectually."

In a foreword to Watson's book Sir Lawrence Bragg praises its "Pepys-like frankness," omitting the not inconsiderable fact that Pepys did not publish his diaries; they were first printed more than a hundred years after his death. Reticence has not been absent from the minds of many as they set out to write accounts of their lives. Thus Edward Gibbon, starting his memoirs:

My own amusement is my motive and will be my reward; and, if these sheets are communicated to some discreet and indulgent friends, they will be secreted from the public eye till the author shall be removed beyond the reach of criticism or ridicule.

But less discreet contemporaries would probably have been delighted had there been a book in which Galilei said nasty things about Kepler. Most things in Watson's book are, of course, not exactly nasty—except perhaps the treatment accorded the late Rosalind Franklin—and some are quite funny, for instance, the description of Sir Lawrence's futile attempts to escape Crick's armor-piercing voice and laughter. It is a great pity that the double helix was not discovered ten years earlier: some of the episodes could have been brought to the screen splendidly by the Marx brothers.

As we read about John and Peter, Francis and Herman, Rosy, Odile, Elizabeth, Linus, and Max and Maurice, we may often get the impression that we are made to look through a keyhole at scenes with which we have no business. This is perhaps unavoidable in an autobiography; but then the intensity of vision must redeem the banality of content. This requirement can hardly be said to be met by Watson's book, which may, however, have a strong coterie appeal, as our sciences are dominated more than ever by multiple cliques. Some of those will undoubtedly be interested in a book in which so many names, and usually first names, appear that are known to them.

This is then a scientific autobiography; and to the extent that it is nothing else, it belongs to a most awkward literary genre. If the difficulties facing a man trying to record his life are great—and few have overcome them successfully—they are compounded in the case of scientists, of whom many lead monotonous and uneventful lives and who, besides, often do not know how to write. Though I have no profound knowledge of this field, most scientific autobiographies that I have seen give me the impression of having been written for the remainder tables of the bookstores, reaching them almost before they are published. There are, of course, exceptions; but even Darwin and his circle come to life much more convincingly in Mrs. Raverat's charming recollections of a Cambridge childhood than in his own autobiography, remarkable a book though it is. When Darwin, hypochondriacally wrapped in his shivering plaid, wrote his memoirs, he was in the last years of his life. This touches on another characteristic facet: scientists write their life's history usually after they have retired from active life, in the solemn moment when they feel that they have not much else to say. This is what makes these books so sad to read: the eagerness has gone: the beaverness remains. In this respect, Watson's book is quite exceptional: when it begins he is 23, and 25 when it ends; and it was written by a man not yet 40.

There may also be profounder reasons for the general triteness of scientific autobiographies. *Timon of Athens* could not have been written, *Les De-*

moiselles d'Avignon not have been painted, had Shakespeare and Picasso not existed. But of how many scientific achievements can this be claimed? One could almost say that, with very few exceptions, it is not the men that make science; it is science that makes the men. What A does today, B or C or D could surely do tomorrow.

Hence the feverish and unscrupulous haste that Watson's book reflects on nearly every page. On page 4: "Then DNA was still a mystery, up for grabs, and no one was sure who would get it and whether he would deserve it. . . . But now the race was over and, as one of the winners, I knew the tale was not simple. . . ." And on page 184: "I explained how I was racing Peter's father [Pauling] for the Nobel Prize." Again on page 199: "I had probably beaten Pauling to the gate." These are just a few of many similar instances. I know of no other document in which the degradation of present-day science to a spectator sport is so clearly brought out. On almost every page, you can see the protagonists racing through the palaestra, as if they were chased by the Hound of Heaven—a Hound of Heaven with a Swedish accent.

There were, of course, good reasons for the hurry, for these long-distance runners were far from lonely. They carried, however, considerably less baggage than others whom they considered, sometimes probably quite wrongly, as their competitors. Quite a bit was known about DNA: the discovery of the base-pairing regularities pointed to a dual structure; the impact of Pauling's α-helix prepared the mind for the interpretation of the x-ray data produced by Wilkins, Franklin, and their collaborators at King's College without which, of course, no structural formulation was possible. The workers at King's College, and especially Miss Franklin, were naturally reluctant to slake the Cavendish couple's thirst for other people's knowledge, before they themselves had had time to consider the meaning of their findings. The evidence found its way, however, to Cambridge. One passage must be quoted. Watson goes to see the (rather poor) film *Ecstasy* (p. 181):

Even during good films I found it almost impossible to forget the bases. The fact that we had at last produced a stereochemically reasonable configuration for the backbone was always in the back of my head. Moreover, there was no longer any fear that it would be incompatible with the experimental data. By then it had been checked out with Rosy's precise measure-ments. Rosy, of course, did not directly give us her data. For that matter, no one at King's realized they were in our hands. We came upon them because of Max's membership on a committee appointed by the Medical Research Council to look into the research activities of Randall's lab. Since Randall wished to convince the outside committee that he had a productive research group, he had instructed his people to draw up a comprehensive summary of their accomplishments. In due time this was prepared in mimeograph form and sent routinely to all the committee members. As soon as Max saw the sections by Rosy and Maurice, he brought the report in to Francis and me. Quickly scanning its contents, Francis sensed with relief that following my return from King's I had correctly reported to him the essential features of the B pattern. Thus only minor modifications were necessary in our backbone configuration.

Rosy is Rosalind Franklin, Max stands for Perutz.

As can be gathered from this astonishing paragraph, Watson's book is quite frank. Without indulging in excesses of self-laceration, he is not a "stuffed shirt" and seems to tell what he considers the truth, at any rate, so far as it concerns the others. In many respects, this book is less a scientific autobiography than a document that should be of interest to a sociologist or a psychologist, who could give an assessment that I am not able to supply. Such an analysis would also have to take account of the merciless persiflage concerning "Rosy" (not redeemed by a cloying epilogue) which goes on throughout the book. I knew Miss Franklin personally, as I have known almost all the others appearing in this book; she was a good scientist and made crucial contributions to the understanding of the structure of DNA. A careful reading even of this book will bear this out.

It is perhaps not realized generally to what extent the "heroes" of Watson's book represent a new kind of scientist, and one that could hardly have been thought of before science became a mass occupation, subject to, and forming part of, all the vulgarities of the communications media. These scientists resemble what Ortega y Gasset once called *the vertical invaders*, appearing on the scene through a trap door, as it were. "He [Crick] could claim no clear-cut intellectual achievements, and he was still without his Ph.D." "Already for thirty-five years he [Crick] had not stopped talking and almost nothing of fundamental value had emerged." I believe it is only recently that such terms as the stunt or the scoop have entered the vocabulary of scientists, who also were not in the habit before of referring to each other as smart cookies. But now, the modern version of King Midas has become all too familiar: whatever he touches turns into a publicity release. Under these circumstances, is it a wonder that what is produced may resemble a Horatio Alger story, but will not be a *Sidereus Nuncius*? To the extent, however, that Watson's book may contribute to the much-needed demythologization of modern science, it is to be welcomed.

ERWIN CHARGAFF
*Department of Biochemistry,
Columbia University, New York City*

Alaska: The Measureless Wealth

Glacier Bay. The Land and the Silence. DAVE BOHN. DAVID BROWER, Ed. Sierra Club, San Francisco, 1967. 165 pp., illus. $25.

In *Glacier Bay*, the Sierra Club once again turns to the task of stimulating public awareness of the natural world and of imparting respect for the land. This magnificently illustrated and sensitively written volume, along with such earlier Sierra Club books as those on the Grand Canyon, the Big Sur coast, and the High Sierra, allow one to *see* and to marvel.

The wondrous scenes these volumes contain are themselves the best of all arguments for resisting needless encroachment on them by the mining companies, the loggers, and the dam builders. Although economic analysis is becoming increasingly useful in shaping policy on the use and conservation of natural resources, economists know no way to make benefit-cost analysis adequately reflect the intangible values of wilderness and other natural environments. A view of, say, the Grand Canyon's inner gorge is indisputably of value, but it is not a marketable masterpiece to be sold at auction. Indeed, to put a price on such a scene is to play into the hands of those who would plug the gorge with concrete and flood it. In the realm of benefit-cost analysis, as in the marketplace, the demand is not for abstractions but for ready coin.

Although some of them are keenly appreciative of natural values, economists seem not to have had much suc-

DNA Helix

I recently came across Dr. E. Chargaff's review (1) of J. D. Watson's book *The Double Helix* (2). I was disturbed by his quotation of an episode which relates how I handed to Watson and Crick an allegedly confidential report by Professor J. T. Randall with vital information about the x-ray diffraction pattern of DNA.

As this might indicate a breach of faith on my part, I have tried to discover what historical accuracy there is in Watson's version of the story, which reads as follows (3):

Even during good films I found it almost impossible to forget the bases. The fact that we had at last produced a stereochemically reasonable configuration for the backbone was always at the back of my head. Moreover, there was no longer any fear that it would be incompatible with the experimental data. By then it had been checked out with Rosy's precise measurements. Rosy, of course, did not directly give her data. For that matter, no one at King's realized they were in our hands. We came upon them because of Max's membership on a committee appointed by the Medical Research Council to look into the research activities of Randall's lab. Since Randall wished to convince the outside committee that he had a productive research group, he had instructed his people to draw up a comprehensive summary of their accomplishments. In due time this was prepared in mimeographed form and sent routinely to all committee members. As soon as Max saw the sections by Rosy and Maurice, he brought the report in to Francis and me. Quickly scanning its contents Francis sensed with relief that following my return from King's I had correctly reported to him the essential features of the "B" pattern. Thus only minor modifications were necessary in our backbone configuration.

Watson showed me his book twice in manuscript; I regret that I failed to notice how this passage would be interpreted by others and did not ask him to alter it. The incident, as told by Watson, does an injustice to the history of one of the greatest discoveries of the century. It pictures Wilkins and Miss Franklin jealously trying to keep their data secret, and Watson and Crick getting hold of them in an underhand way, through a confidential report passed on by me. What historical evidence I have been able to collect does not corroborate this story. In summary, the committee of which I was a member did not exist to "look into the research activities of Randall's lab," but to bring the different Medical Research Council units working in the field of biophysics into touch with each other. The report was not confidential and contained no data that Watson had not already heard about from Miss Franklin and Wilkins themselves. It did contain one important piece of crystallographic information useful to Crick; however, Crick might have had this more than a year earlier if Watson had taken notes at a seminar given by Miss Franklin.

I discarded the papers of the committee many years ago but the Medical Research Council kindly found them for me in their archives. According to their records there were, in fact, two committees. First, the Biophysics Research Unit Advisory Committee, set up at the beginning of 1947 "to advise regarding the scheme of research in biophysics under the direction of Professor J. T. Randall." Neither Randall nor I were members of that committee; I did not know of its existence until recently. It held its final meeting in October 1947, 5 years before the episode related by Watson. Later that year the Council set up the Biophysics Committee "to advise and assist the Council in promoting research work over the whole field of biophysics in relation to medicine." This new committee consisted mainly of the heads of all the Medical Research Council units related to biophysics, and included Randall and myself. We visited each laboratory in turn; the director would tell the others about the research in his unit and circulate a report. The reports were not confidential. The committee served to exchange information but was not a review body; we were never asked for an opinion of the work we saw. The Medical Research Council dissolved it in 1954, in the words of the official letter because "the Committee has fulfilled the purpose for which it was set up, namely to establish contact between the groups of people working for the Council in this field" (Appendix 1).

On 15 December 1952, we met in Randall's laboratory where he gave us a talk and also circulated the report referred to in Watson's book. As far as I can remember, Crick heard about its existence from Wilkins, with whom he had frequent contact, and either he or Watson asked me if they could see it. I realized later that, as a matter of courtesy, I should have asked Randall for permission to show it to Watson and Crick, but in 1953 I was inexperienced and casual in administrative matters and, since the report was not confidential, I saw no reason for withholding it.

I now come to the technical details of the report. It includes one short section describing Wilkins' work on DNA and nucleoprotein structures and then another on "X-ray studies of calf thymus DNA" by R. E. Franklin and R. G. Gosling. They are reproduced in Appendix 2 below. Note that they contain only two pieces of numerical data. One is the length of the fiber axis repeat of 34 Å in the wet or "B" form of DNA; this is the biologically more important form, solved by Watson and Crick. The other piece consists of the unit-cell dimensions and symmetry of the partially dried "A" form, which was the one discovered and worked on by Wilkins and Miss Franklin, to be solved later by Wilkins and his colleagues. The report contained no copies of the x-ray diffraction photographs of either form.

We can now ask if this section really contained "Rosy's precise measurements needed to check out" Watson and Crick's tentative model and whether it is true that "Rosy did not give us her data . . . and no one at King's realized that they were in our hands." In fact, the report contained no details of the vital "B" pattern apart from the 34 Å repeat, but Watson, according to his own account heard them from Wilkins himself, shortly before he saw the report. This story is told in chapter 23, relating Watson's visit to King's College in late January 1953 where Miss Franklin supposedly tried to hit him and where Wilkins showed him a print of one of her exciting new x-ray photographs of the "B" form of DNA. The next chapter (24) begins as follows: "Bragg was in Max's office when I rushed in the next day to blurt out what I had learned. Francis was not yet in, for it was a Saturday morning and he was home in bed glancing at the *Nature* that had come in the morning mail. Quickly I started to run through the details of the "B" form of DNA, making a rough sketch to show the evidence that DNA was a helix which repeated its pattern every 34 Å along the helical axis." The incident of the report comes in the following chapter (25) and is dated early 1953.

It is interesting that a drawing of the "B" patterns from squid sperm is also contained in a letter from Wilkins to Crick written before Christmas 1952. All this clearly shows that Wilkins disclosed many, even though perhaps not all, of the data obtained at King's to either Watson or Crick.

It took over a year for Max Perutz to orchestrate his, Wilkins', and Watson's responses to Chargaff's review. Their letters were published in *Science*, June 27, 1969.

Turning now to the x-ray pattern of the "A" form, this had been the subject of a seminar given by Miss Franklin at King's in November 1951, an occasion described by Watson in chapter 10. After Miss Franklin's tragic death in 1958, her colleague, Dr. A. Klug, preserved her scientific papers; among these are her notes for that seminar, which he now kindly showed me. These notes include the unit-cell dimensions and symmetry of the "A" form which were circulated in the report a year later.

Watson, according to his own account, had failed to take notes at Miss Franklin's seminar, so that he could not give the unit-cell dimensions and symmetry to Crick afterward. Crick tells me now that the report did bring the monoclinic symmetry of the unit cell home to him for the first time. This really was an important clue as it suggested the existence of twofold symmetry axes running normal to the fiber axis, requiring the two chains of a double helical model to run in opposite directions, but he could clearly have had this clue much earlier.

MAX F. PERUTZ

42 Sedley Taylor Road,
Cambridge, England

References and Notes

1. E. Chargaff, *Science* **159**, 1448 (1968).
2. J. D. Watson, *The Double Helix, A Personal Account of the Discovery of the Structure of DNA* (Athenuem, New York, 1968).
3. ———, *ibid.,* p. 181.
4. I thank the Medical Research Council, Dr. A. Klug, and Dr. R. Olby for supplying me with historical documents, and Sir J. Randall, Professor M. H. F. Wilkins, and Dr. R. G. Gosling for permission to publish their report.

10 April 1969

Appendix 1

27 April 1954

Dear Perutz

The Council have been considering the future of their Biophysics Committee, which was appointed in 1947 and would be due for reconstitution if it were to be kept in being. After consultation with the Chairman and others, they have come to the conclusion that *the Committee has fulfilled the purpose for which it was set up, namely to establish contact between the different groups of people working for the Council in this field.* It has accordingly been decided that the Committee should now be discharged. I am asked by the Council to send you their best thanks for all the help that you have given to their work by serving on this Committee.

Yours sincerely,
Landsborough Thomson
(Secretary to the Biophysics Committee)

Appendix 2

Report by Professor J. T. Randall to the Medical Research Council, dated December 1952

Nucleic Acid Research

The research on nucleic acids, like that on collagen, has both a structural and a biological interest. Some time ago Wilkins found that fibres from sodium desoxyribonucleate gave remarkably good x-ray fibre diagrams. He also examined the optical properties of the fibres in relation to their molecular structure. The detailed examination of the structure has been continued by Miss Franklin and R. G. Gosling, and Wilkins has concentrated on a study of the oriented nucleoprotein of sperm heads. The biological implications of this work are indicated later in this section.

The study of nucleic acids in living cells has been continued by Walker (tissue cultures) and by Chayen (plant root meristem cells); and lately Wilkins and Davies have been measuring the dry weight of material in *Tradescantia* pollen grains during the course of cell division by means of interference microscopy. Thus, while the work of Walker on nucleic acid content of nuclei relates only to part of the cell contents, the interference microscope enables the total content of the cell, other than water, to be measured.

Desoxyribose Nucleic Acid and Nucleoprotein Structure (M. H. F. Wilkins)

A molecular structure approach has been made to the question of the function of nucleic acid in cells.

First, x-ray evidence shows that DNA from all kinds of sources has the same basic molecular configuration which is little (if at all) dependent on the nucleotide ratio. Some grouping of polynucleotide chains takes place to give ~ 20 Å diameter rod-shaped units, and the internal chemical binding which holds each unit together is not affected much by the normal extraction procedure. The basic point is to find the general nature of this structure and the hydrogen bonding etc. in it. Using two dimensional data, the most reasonable interpretation was in terms of a helical structure and the experimental evidence for such helices was much clearer than that obtained for any protein. The crystalline material gives an x-ray picture with considerable elements of simplicity which could be accounted for by the helical ideas, but three dimensional data show apparently that the basic physical explanation of the simplicity of the picture lies in some quite different and, a priori, much less likely structural characteristic. The 20 Å units, while roughly round in cross-section, appear to have highly asymmetric internal structure.

The same general configuration appears to exist in intact sperm heads and synthetic or extracted nucleoprotein, and in bacteriophage (and not in insect virus

where the protein is different). It appears that the protein is probably bound electrostatically on the outside of the nucleic acid units and does not alter their structure. In some sperm the whole head has a crystalline (but somewhat imperfect) structure. In these sperm, the protein has very low molecular weight and it will be especially interesting to find if any high molecular weight protein exists in such sperm heads. If not, all the genetical characteristics may be supposed to lie in the DNA (as in bacteriophage). Biochemical study of the composition of the protein is planned. In other kinds of cell nucleus with different biological function the proteins are quite different. The main idea is to find the structure of the DNA first, then how it is linked to protein in the crystalline sperm heads, and then attempt to elucidate the more complex structure of the other kinds of cell nuclei. It may be that the characteristic x-ray picture of DNA is especially related to a particular function of the nuclear nucleoprotein. In this way molecular structure and cytochemical studies begin to overlap.

X-ray Studies of Calf Thymus DNA (R. E. Franklin and R. G. Gosling)

(a) The Role of Water: The crystalline form of calf thymus DNA is obtained at about 75 percent RH and contains about 20 percent by weight of water.

Increasing the water content leads to the formation of a different structural modification which is less highly ordered. The water content of this form is ill-defined.

The change from the first to the second structure is accompanied by a change in the fibre-axis repeat period of 28 Å to 34 Å and a corresponding microscopic length-change of the fibre of about 20 percent.

Decreasing the water-content below 20 percent leads to a gradual fading out of the crystalline x-ray pattern and a corresponding increase in the diffuse background scattering. After strong drying only diffuse scattering is observed.

All these changes are readily reversible. The following explanation is suggested:

The phosphate groups, being the most polar part of the structure would be expected to associate with one another and also with the water molecules. Phosphate-phosphate bonds are considered to be responsible for intermolecular linking in the crystalline structure. The water molecules are grouped around these bonds (approximately four water molecules per phosphorus atom). Increased water content weakens these bonds and leads, first, to a less highly ordered structure and, ultimately, to gel formation and solution. Drying leaves the phosphate-phosphate links intact but leads to the formation of holes in the structure with resulting strain and deformation. The three-dimensional skeleton is preserved in distorted form and crystalline order is restored when the humidity is again increased.

(b) The Cylindrically Symmetrical Patterson Function: It was apparent that the

crystalline form was based on a face-centered monoclinic unit cell with the *c*-axis parallel to the fibre axis. But it was not found possible, by direct inspection, to allot all the lattice parameters accurately and unambiguously. To obtain the unit cell with certainty the cylindrically symmetrical Patterson function was calculated. This function is periodic in the fibre-axis direction only.

Special techniques were developed for the measurement of the positions and intensities of the reflections. This was necessary, firstly because all measurements had to be made on micro-photographs, and secondly because the observed reflections were of a variety of shapes and sizes so that integrated intensities could not be directly measured.

On the Patterson function obtained, the lattice translations could be readily identified. On the basis of a unit cell defined by

$$a = 22.0 \text{ Å}$$
$$b = 39.8 \text{ Å}$$
$$c = 28.1 \text{ Å}$$
$$\beta = 96.5°$$

the 66 independent reflections observed could all be indexed with an error of less than 1 percent.

A very satisfactory confirmation of the correctness of the unit cell and the indexing was provided by a fortunate accident which it has so far not been possible to reproduce. One fibre was obtained which gave a photograph showing strong double orientation. It was found that in this photograph those spots which had been indexed *hkl* were strongest in one pair of quadrants while those indexed *hkl* were strongest in the other pair.

(c) *The Three-Dimensional Patterson Function*: Having established the unit cell with certainty, it is now possible to calculate Patterson sections in the normal way. Work on these is in progress.

In Dr. M. F. Perutz's letter, extracts from a Medical Research Council report are published for the first time. For those interested in the history of the early x-ray studies of DNA at King's college, I give here the main facts which form the background to the report.

Early in 1951 "A" patterns of DNA and very diffuse "B" patterns from DNA and from sperm heads indicated (as I described at a meeting at Cambridge in 1951) that DNA was helical. Shortly afterward, when Rosalind Franklin began experimental work on DNA, she almost immediately obtained (in September 1951) the first clear "B" patterns [described at a seminar in 1951 and published in 1953 (*1*)]. By the beginning of 1952 I had obtained basically similar patterns from DNA from various sources and from sperm heads. The resemblance (*2*) of the "B" patterns of DNA and those of

sperm was very clear at that time. The helical interpretation was very obvious too, and it was proposed in general terms in Franklin's fellowship report (*3*). The "B" patterns of DNA that I obtained at that time were quite adequate for a detailed helical interpretation. This was given later (*4*), with one of the patterns, alongside the Watson and Crick description (*5*) of their model. The best, and most helical-looking "B" pattern, was obtained by Franklin in the first half of 1952 and was published in 1953 (*6*), also with a helical interpretation and alongside the Watson-Crick paper. Confusion arose because, during the summer of 1952, Franklin presented, in our laboratory, "A"-type data (in three dimensions) which showed that the DNA molecule was asymmetrical and therefore nonhelical. Later in the year I wrote for the Medical Research Council report a summary of the DNA x-ray work as a whole in our laboratory. Since our previous emphasis had been entirely on helices, I drew attention in the report to the nonhelical interpretation. In 1953, after the Watson-Crick model had been built and when we had more precise "A" data, I reexamined the question of DNA being nonhelical and found that the data gave no support for the molecule being nonhelical (*7*).

M. H. F. WILKINS

Medical Research Council,
Biophysics Research Unit,
King's College, London

References

1. R. E. Franklin and R. G. Gosling, *Acta Cryst.* **6**, 673 (1953).
2. M. H. F. Wilkins and J. T. Randall, *Biochim. Biophys. Acta* **10**, 192 (1953).
3. A. Klug, *Nature* **219**, 808 (1968).
4. M. H. F. Wilkins, A. R. Stokes, H. R. Wilson, *ibid* **171**, 738 (1953).
5. J. D. Watson and F. H. C. Crick, *ibid.*, p. 737.
6. R. E. Franklin and R. G. Gosling, *ibid.*, p. 740.
7. M. H. F. Wilkins, W. E. Seeds, A. R. Stokes, H. R. Wilson, *ibid.* **172**, 759 (1953).

10 April 1969; revised 26 May 1969

I am very sorry that, by not pointing out that the Randall report was nonconfidential, I portrayed Max Perutz in a way which allowed your reviewer [*Science* **159**, 1448 (1968)] to badly misconstrue his actions. The report was never marked "confidential," and I should have made the point clear in my text [*The Double Helix* (Athenum, New York, 1968)]. It was my intention to reconstruct the story accurately, and so most people mentioned in the story were given the manuscript, either in first draft or in one of the subsequent

revisions, and asked for their detailed comments.

I must also make the following comments.

1) While I was at Cambridge (1951–53) I was led to believe by general lab gossip that the MRC (Medical Research Council) Biophysics Committee's real function was to oversee the MRC–King's College effort, then its biggest venture into pure science. I regret that Perutz did not ask me to change this point.

2) The Randall report was really very useful, especially to Francis [Crick]. In writing the book I often underdescribed the science involved, since a full description would kill the book for the general reader. So I did not emphasize, on page 181, the difference between "A" and "B" patterns. The relevant fact is not that in November 1951 I *could have* copied down Rosalind's seminar data on the unit cell dimensions and symmetry, but that I *did not*. When Francis was rereading the report, after we realized the significance of the base pairs and were building a model for the "B" structure, he suddenly appreciated the diad axis and its implication for a two-chained structure. Also, the report's explicit mention of the "B" form and its obvious relation to the expansion of DNA fiber length with increase of the surrounding humidity was a relief to Francis, who disliked my habit of never writing anything on paper which I hear at meetings or from friends. The fiasco of November 1951 arose largely from my misinterpretation of Rosy's talk, and with my knowledge of crystallography not really much solider, I might have easily been mistaken again. Thus the report, while not necessary, was very, very helpful. And if Max had not been a member of the committee, I feel that neither Francis nor I would have seen the report; and so, it was a fluke that we saw it.

3) Lastly, Max's implication that the King's lab was generally open with all their data badly oversimplifies a situation which, in my book, I attempted to show was highly complicated in very human ways.

All these points aside, I regret and apologize to Perutz for the unfortunate passage.

JAMES D. WATSON

The Biological Laboratories,
Harvard University,
Cambridge, Massachusetts

19 May 1969

Acknowledgments

A glance at any page in this book reveals the great debt we owe to many individuals and archives throughout the world.

Jim Watson encouraged us to undertake the project and left us alone to get on with it. Seeing the book in rough page proofs reawakened his memories, and he provided us with new leads to investigate. Ray Gosling has been most generous with his time, memories, and comments. We are particularly grateful for the pieces he wrote specially for the book. These provide insights into the work that he did with Maurice Wilkins and Rosalind Franklin and help us appreciate the dynamics of the MRC Biophysics Unit at King's College, London. It was a great pleasure to contact others who were also part of the story. Bruce Fraser, Herbert ("Freddie") Gutfreund, Hugh Huxley, Av Mitchison, and Alex Rich provided photographs. They are evocative of the period and provide visual counterparts to Watson's descriptions. After much detective work we found Bertrand Fourcade in Paris and had a delightful conversation with him. We are very grateful to Michael Crick for providing the wonderful letter written to him by his father, describing the discovery of the double helix and its significance. Jenifer Glynn, Rosalind Franklin's sister, very kindly directed us to her sister's papers in the Ann Sayre Archive at the University of Maryland, and in the Churchill College Archive Centre, Cambridge.

Angela Creager, Ray Gosling, Walter Gratzer, and Robert Olby read the manuscript and their comments and suggestions were extremely useful in helping us refine our annotations.

Our project would not have been possible without the archivists whose hard work cataloging and digitizing documents has been invaluable.

The papers of Jim Watson and Sydney Brenner are at the Cold Spring Harbor Laboratory Archive (directed by Ludmila Pollock) and we are very grateful to John Zarillo, whose knowledge of the Watson Archive is second to none.

John was endlessly patient with our numerous requests for information and images, and prompt in responding to all of them, however obscure. Christopher Olver was very helpful when he was cataloging the Maurice Wilkins Papers at King's College, London, directed by Geoff Browell. The Crick papers are held by the Wellcome Trust, and Jennifer Haynes and Helen Wakely in Archives and Manuscripts of the Wellcome Library (directed by Simon Chaplin) sent us materials and information. Max Delbrück was a regular correspondent of Watson and so it was especially valuable that Shelley Erwin (Head of Archives) and Loma Karklins of Archives and Special Collections, California Institute of Technology, sent us the letters which Delbrück received from Watson during his time in Cambridge. We thank Chris Petersen of the Special Collections & Archives Research Center at Oregon State University for all his help. The Linus and Ava Pauling Archives are an extraordinary resource for materials relating to the period we cover and the online materials greatly helped us with Pauling's part in the story. There is little of Rosalind Franklin's correspondence in this period, primarily because, living in London, she visited her parents rather than write to them. However, Anne Sayre deposited the materials she used in writing *Rosalind Franklin & DNA* with the archive of the American Society of Microbiology at the University of Maryland, Baltimore Campus. We are very grateful to Jeff Karr for being our guide. The Archive Centre at Churchill College, Cambridge, holds, among others, papers of Rosalind Franklin and John Randall. We are grateful to Alan Packwood (Director) and Sophie Bridges for their assistance.

Many others provided information and illustrations, acknowledged individually in the sources and list of figures. These include:

Enrico Alleva and Claudia Di Somma (Stazione Zoologica Anton Dohrn); Liz Allsopp and Maggie Johnston (Rothamsted Research, United Kingdom); Liz Bass (Stony Brook University); Richard Beyler (Portland State University); Janet Browne (Harvard University); Maurizio Brunori (University of Rome); Donald Caspar (Florida State University); Jean-Pierre Changeux (Institut Pasteur); John Collinge (University of London); Peter Collins (Royal Society, London); Helen Drury (Oxfordshire County Council); Jack Dunitz (Swiss Federal Institute of Technology);

Jose Elguero (Instituto de Quimica Medica, Spain); George Elliott (King's College, London); Annette Faux (MRC Laboratory of Molecular Biology, Cambridge); Jonathan Ford (Athenaeum Club, London); Jenifer Glynn; Janice F. Goldblum (National Academy of Sciences Archives); Tom Hager (Oregon State University); Kersten Hall, Adam Nelson, and Bruce Turnbull (University of Leeds); Stephen Harding (University of Nottingham); Ken Holmes (MPIMF Heidelberg); Gareth Jones (King's College, London); John Krige (Georgia Institute of Technology); John Lagnardo; Debra Lyons and Gill Shapland (Cambridgshire Archives); Kristie Macrakis (Georgia Institute of Technology); Brenda Maddox; Neil Mitchison (European Commission Office, Scotland); Val Mitchison; Luke O'Neill (Trinity College, Dublin); Gail Oskin (Harvard University); Mike Petty; Jeffrey Reznick and Paul Theerman (National Library of Medicine); Daniel Salsbury (National Academy of Sciences); Don Stone (University of Wisconsin, Madison); Gary Todoroff (North Coast Photos, Eureka, California); Inder Verma (Salk Institute); Judy Wilson (Cambridge, United Kingdom); Michael Woolfson (University of York); and Anne Cabot Wyman (Cambridge, Massachusetts)

We would like to acknowledge Alfred A. Knopf, a division of Random House Inc., and Oxford University Press who gave us permission to reprint the chapter excerpt from *Avoiding Boring People*. We thank Orion for permission to distribute this edition in the UK and associated territories.

We have used a very large number of illustrations and we thank the following institutions who gave permission to use the figures and waived or reduced their charges:

The Athenaeum Club, London; California Institute of Technology; Cambridge University Archives; Churchill College, Cambridge; King's College, London; MRC Laboratory for Molecular Biology, Cambridge; Medical Research Council, London; *Nature*; Oregon State University; *Proceedings of the National Academy of Sciences*; *Science*; Smithsonian Institution; University of California, San Diego; Wellcome Trust; John Wiley.

We thank the team at CSHL Press. This was an unusual production in which, because the annotations and illustrations were to be closely tied to the original text, we went to layouts immediately. That we were able to do so was due to Denise Weiss' page design which was immediately right and to her continuing oversight of the design. Susan Schaefer took our collections of notes and figures and transformed them into proper pages. She was indefatigable in coping with an almost never ending stream of changes. We thank Carol Brown and Inez Saliano who undertook the daunting task of getting permissions for all the material. Without their work the book would be much diminished. We are indebted to Jan Argentine (Director, Book Development) and John Inglis (Publisher, CSHL Press) for their skilled and enthusiastic management of the project, without which our writing lives would have been much more difficult.

We are grateful to Amanda Urban of ICM Partners for facilitating our collaboration with Simon & Schuster, and to Jonathan Karp (Publisher) and the team of Karyn Marcus (Senior Editor), Irene Kheradi (Managing Editor), her associate Gina DiMascia, and Michael Accordino (Art Director), which has moved at lightning speed to bring the book to press.

Bibliography

Andrew C. 2009. *Defend the realm: The authorized history of MI5.* Alfred A. Knopf, New York.

Appel TA. 2000. *Shaping biology: The National Science Foundation and American Biological Research, 1945–1975.* Johns Hopkins University Press, Baltimore.

Benton J. 1990. *Naomi Mitchison: A biography.* Pandora, London.

Berg P, Singer M. 2003. *George Beadle: An uncommon farmer.* Cold Spring Harbor Laboratory Press, Cold Spring Harbor, NY.

Bernal JD. 1939. *The social function of science.* Routledge & Kegan Paul Ltd., London.

Beskow B. 1969. *Dag Hammarskjöld: Strictly personal: A portrait.* Doubleday, New York.

Binder M. 2011. *Halliwell's horizon: Leslie Halliwell and his film guides.* lulu.com.

Brown A. 2005. *J.D. Bernal: The sage of science.* Oxford University Press, Oxford.

Bullard M. 1952. *A perch in paradise.* Hamish Hamilton, London.

Cairns J, Stent GS, Watson JD, eds. 1992. *Phage and the origins of molecular biology,* expanded edition. Cold Spring Harbor Laboratory Press, Cold Spring Harbor, NY.

Carlson EA. 1981. *Genes, radiation, and society: The life and work of H.J. Muller.* Cornell University Press, Ithaca, NY.

Chargaff E. 1963. *Essays on nucleic acids.* Elsevier, Amsterdam.

Chargaff E. 1978. *Heraclitean fire: Sketches from a life before nature.* The Rockefeller University Press, New York.

Chomet S, ed. 1995. *DNA: Genesis of a discovery.* Newman-Hemisphere Press, London.

Coyne JS. 1855. *Pippins & pies, or, sketches out of school: Being the adventures and misadventures of Master Frank Pickleberry during that month he was home for the holidays.* G. Routledge & Co., London.

Creager ANH. 2002. *The life of a virus: Tobacco mosaic virus as an experimental model, 1930–1965.* University of Chicago Press, Chicago.

Crick FHC. 1988. *What mad pursuit: A personal view of scientific discovery.* Basic Books, New York.

Crowther JG. 1974. *The Cavendish Laboratory, 1874–1974.* Macmillan, New York.

Davidson JN. 1950. *The biochemistry of the nucleic acids.* Methuen & Co. Ltd., London.

de Chadarevian S. 2002. *Designs for life: Molecular biology after World War II.*

Cambridge University Press, Cambridge.

Dubos RJ. 1976. *The professor, the institute, and DNA. Oswald T. Avery: His life and scientific achievements.* The Rockefeller University Press, New York.

Eden RJ. 1998. *Clare College and the founding of Clare Hall.* Clare Hall, Cambridge.

Edwards P. 2000. *Wyndham Lewis: Painter and writer.* Yale University Press, New Haven.

Ellicot D. 1975. *Our Gibraltar: A short history of the rock.* Gibraltar Museum, Gibraltar.

Fell HB. 1962. *History of the Strangeways Research Laboratory (formerly Cambridge Research Hospital) 1912–1962.* Strangeways Research Laboratory, Cambridge.

Ferry G. 1998. *Dorothy Hodgkin: A life.* Granta Books, London.

Ferry G. 2007. *Max Perutz and the secret of life.* Cold Spring Harbor Laboratory Press, Cold Spring Harbor, NY.

Finch J. 2008. *A Nobel fellow on every floor: A history of the Medical Research Council Laboratory of Molecular Biology.* MRC Laboratory of Molecular Biology, Cambridge.

Fischer EP, Lipson C. 1988. *Thinking about science: Max Delbrück and the origins of molecular biology.* W.W. Norton, New York.

Friedberg EC. 2005. *The writing life of James D. Watson.* Cold Spring Harbor Laboratory Press, Cold Spring Harbor, NY.

Gamow G. 1970. *My world line: An informal autobiography.* Viking Press, New York.

Glynn J. 2012. *My sister Rosalind Franklin.* Oxford University Press, Oxford.

Goldwasser E. 2011. *A bloody long journey: Erythropoietin (Epo) and the person who isolated it.* Xlibris Corporation, Dartford, UK.

Hager T. 1995. *Force of nature: The life of Linus Pauling.* Simon & Schuster, New York.

Hoffmann D, Ehlers J, Renn J, eds. 2007. *Pascual Jordan (1902–1980). Mainzer Symposium zum 100. Geburtstag,* preprint 329. Max Planck Institute for the History of Science, Berlin.

Inglis J, Sambrook J, Witkowski J, eds. 2003. *Inspiring science: Jim Watson and the age of DNA.* Cold Spring Harbor Laboratory Press, Cold Spring Harbor, NY.

Jacob F. 1988. *The statue within.* Basic Books, New York.

Jones RV. 1978. *Most secret war: British scientific intelligence 1939–1945.* Hamish Hamilton, London.

Judson HF. 1996. *The eighth day of creation: The makers of the revolution in biology.* Cold Spring Harbor Laboratory Press, Cold Spring Harbor, NY.

Kamminga H, de Chadarevian S. 2002. *Representations of the double helix*. Whipple Museum of the History of Science, Cambridge.

Kilmister CW, ed. 1987. *Schrödinger: Centenary celebration of a polymath*. Cambridge University Press, Cambridge.

Kohler RE. 1991. *Partners in science: Foundations and natural scientists, 1900–1945*. University of Chicago Press, Chicago.

Lessing D. 1997. *Walking in the shade: Volume two of my autobiography, 1949 to 1962*. Harper Collins, London.

Luria SE. 1984. *A slot machine, a broken test tube: An autobiography*. Harper & Row, New York.

Maddox B. 2002. *Rosalind Franklin: The dark lady of DNA*. Harper Collins, London.

Marage P, Wallenborn G, eds. 1999. *The Solvay Councils and the birth of modern physics*. Birkhäuser, Basel, Switzerland.

Martin A. 2002. *Herring fishermen of Kintrye and Ayshire*. Bell & Bair, Glasgow.

McCarty M. 1990. *The transforming principle: Discovering that genes are made of DNA*. W.W. Norton, New York.

McElheny VK. 2003. *Watson and DNA: Making a scientific revolution*. Perseus Publishing, Cambridge, MA.

McElroy WD, Glass B, eds. 1957. *The chemical basis of heredity*. Johns Hopkins Press, Baltimore.

Monk R. 1990. *Ludwig Wittgenstein: The duty of genius*. The Free Press, New York.

Monod J, Borek E. 1971. *Of microbes and life*. Columbia University Press, New York.

Morton F. 1962. *The Rothschilds: A family portrait*. Atheneum, New York.

Nobel Foundation. 1963. *Nobel lectures: Physiology or medicine, 1942–1962*. Elsevier, Amsterdam.

Norrington ALP. 1983. *Blackwell's 1879–1979: The history of a family firm*. Blackwell Publishers, Oxford.

Olby R. 1994. *The path to the double helix: The discovery of DNA*. Dover, Mineola, NY.

Olby R. 2009. *Francis Crick: Hunter of life's secrets*. Cold Spring Harbor Laboratory Press, Cold Spring Harbor, NY.

Pauling L. 1939. *The nature of the chemical bond and the structure of molecules and crystals*. Cornell University Press, Ithca, NY.

Perutz M. 1997. *Science is not a quiet life: Unravelling the atomic mechanism of haemoglobin*. World Scientific, Singapore.

Perutz M. 2002. *I wish I'd made you angry earlier: Essays on science, scientists, and humanity,* expanded edition. Cold Spring Harbor Laboratory Press, Cold Spring Harbor, NY.

Ray N. 1994. *Cambridge architecture: A concise guide.* Cambridge University Press, Cambridge.

Ridley M. 2006. *Francis Crick: Discoverer of the genetic code.* Harper Collins, London.

Roach JPC, ed. 1959. *A History of the county of Cambridge and the Isle of Ely: Volume 3: The city and University of Cambridge.*

Roberts SC. 2010. *Adventures with authors.* Cambridge University Press, Cambridge.

Rocca J. 2007. *Forging a medical university—The establishment of Sweden's Karolinska Institutet.* Karolinska Institutet Press, Stockholm.

Sayre A. 1975. *Rosalind Franklin and DNA.* W.W. Norton, New York.

Shearer SM. 2010. *Beautiful: The life of Hedy Lamarr.* Thomas Dunne Books, New York.

Soderqvist T. 2003. *Science as autobiography: The troubled life of Niels Jerne.* Yale University Press, New Haven.

Stoff J. 2009. *John F. Kennedy International Airport.* Arcadia Publishing, South Carolina.

Swann B, Aprahamian F. 1999. *J.D. Bernal: A life in science and politics.* Verso, London.

Thomas JM, Phillips D, eds. 1990. *Selections and reflections: The legacy of Sir Lawrence Bragg.* The Royal Institution of Great Britain, London.

Todd A. 1973. *A time to remember: The autobiography of a chemist.* Cambridge University Press, Cambridge.

Watson JD. 1968. *The Double Helix: A personal account of the discovery of the structure of DNA.* Atheneum, New York.

Watson JD. 1980. *The Double Helix,* The Norton Critical Edition (ed. Stent G). W.W. Norton, New York.

Watson JD. 2001. *Genes, girls and Gamow.* Oxford University Press, Oxford.

Watson JD. 2001. *A passion for DNA*: *Genes, genomes, and society.* Cold Spring Harbor Laboratory Press, Cold Spring Harbor, NY.

Watson JD. 2007. *Avoid boring people.* Oxford University Press, Oxford.

Werskey G. 1978. *The visible college: A collective biography of British scientists and socialists of the 1930s.* Allen Lane, London.

Wilkins M. 2003. *The third man of the double helix.* Oxford University Press, Oxford.

Wilson J. 2010. *Cambridge Grocer: The story of Matthew's of Trinity Street 1832–1962.* Bertram, Cambridge.

Witkowski JA. 2000. *Illuminating life: Selected papers from Cold Spring Harbor (1903–1969).* Cold Spring Harbor Laboratory Press, Cold Spring Harbor, NY.

Wyman AC. 2010. *Kipling's cat: A memoir of my father.* Protean Press, Rockport, MA.

Wyman J. 2010. *Alaska Journal 1951.* Protean Press, Rockport, MA.

Wyman J. 2010. *Letters from Japan 1950.* Protean Press, Rockport, MA.

Sources

Full details of books cited by author and date of publication can be found in the bibliography.

Archives

ASM   American Society of Microbiology Archive, University of Maryland, Baltimore Campus.

CALTA   California Institute of Technology Archives.

CAV   Cavendish Laboratory, Department of Physics, Cambridge.

CCAC   Churchill College Archives Centre.

CSHLA   Cold Spring Harbor Laboratory Archives.

KCLA   King's College, London, Archives.

OSUSC   Ava Helen and Linus Pauling Papers, 1873–2011, Oregon State University Special Collections.

WL   Crick, Francis Harry Compton (1916–2004), The Wellcome Library.

Prologue

1. "Willy Seeds' remark... ." William E. Seeds was a colleague of Maurice Wilkins in the Medical Research Council Biophysics Unit at King's College, London. Seeds published papers with Wilkins on light microscopy and on the structure of DNA; Watson, 2001. *A passion for DNA: Genes, genomes, and society*, p. 21.

Chapter 1

1. Crowther, 1974.
2. "Lawrence Bragg... ." da C Andrade EN, Lonsdale K. 1943. William Henry Bragg. 1862–1942. *Obit Not Fell R Soc* **4:** 276–300; Phillips D. 1979. William Lawrence Bragg. 31 March 1890–1 July 1971. *Biogr Mems Fell R Soc* **25:** 74–143.
3. Authors' personal observations.
4. "A caricature of Crick... ." *WL*.
5. "The fellows of a college... ." Cambridge University Senior Tutors' Committee. *The educational provision of the Cambridge Colleges*, 2005.
6. "The current opulence of King's College... ." Roach, 1959, pp. 376–408.

Chapter 2

1. "*What Is Life*? is based… ." Perutz M. 1987. *Erwin Schrödinger's* What is Life? *and molecular biology* (ed. Kilmister CW), pp. 234–251; Crick to Schrödinger, August 12, 1953, the Dublin Institute of Advanced Studies Archive.

2. Avery O, MacLeod CM, McCarty V. 1944. Studies on the chemical nature of the substance inducing transformation of pneumococcal types. Induction of transformation by a desoxyribonucleic acid fraction isolated from pneumococcus type III. *J Exp Med* **79**: 137–159; Dubos, 1976.

3. "Not to be confused with… ." Hearnshaw, 1929.

4. "While Watson writes… ." Maddox, 2002, p. 110; Wilkins, 2003, p. 165.

5. "The nickname 'Rosy'… ." Maddox, 2002, p. 160.

6. "Franklin's claim that DNA… ." Maddox, 2002, pp. 114–115; Wilkins, 2003, pp. 148–150.

7. "While the combination rooms… ." Ray Gosling, BBC Today program, September 30, 2010; Judson, 1996, pp. 625–627.
   "It wasn't just King's… ." Franklin to Anne Sayre, March 1, 1952, *ASM*.

8. "Gerald Oster… ." Oster to Pauling, August 9, 1951; Randall to Pauling, August 28, 1951; Pauling to Randall, September 25, 1951, *OSUSC*.

9. "…as objectionable as physics… ." Wilkins, 2003, p. 83.

Chapter 3

1. "The Naples meeting… ." Struttura Submicroscopia del Protoplasma. Napoli, May 22–25, 1951.

2. "Watson was attracted… ." Watson, 2007, p. 31.

3. "Watson had decided… ." Watson, 2007, p. 45.

4. "Luria and Delbrück met… ." Luria, 1984, pp. 32–35; for essays on the contributions of research on phage to molecular biology, see Cairns et al., 1992.

5. "There were two institutes… ." Witkowski, 2000, p. ix.

6. See Appendix 3 for details.

7. "Kalckar had separated… ." Watson, 2007, pp. 81–82.
   "Watson wrote a long letter… ." Watson to Delbrück, March 22, 1951, *CALTA*.

8. "The Stazione Zoologica Napoli… ." Maienschein J. 1985. First impressions: American biologists at Naples. *Biol Bull* (suppl.) **168**: 187–191.

Chapter 4

1. "John Turton Randall was at… ." Wilkins MHF. 1987. John Turton Randall. March 23, 1905–June 16, 1984. *Biogr Mems Fell R Soc* **33**: 492–535.

2. "Watson's pessimism was justified… ." The papers presented at the meeting were published in *Pubbl Staz Zool Napoli* 23, Supplement **115:** 1–206, 1951. F Buchtal wrote on rheology (p. 115) and RD Preston on *Valonia* (p. 184).
3. "X-ray diffraction photograph… ." Gosling R, personal communication, January 28, 2012.
4. "Watson wasn't the only one… ." Gosling R, personal communication, June 2, 2011.

Chapter 5

1. "In 1948, Pauling was… ." Pauling L. 1955. The stochastic method and the structure of proteins. *Am Sci* **43:** 285–297, pp. 293–294; Hager, 1995, pp. 323–324.
2. "This influence of Pauling's style… ." Watson JD, Crick FHC. 1953. Genetical implications of the structure of deoxyribonucleic acid. *Nature* **171:** 964–967.
3. "In a letter… ." Luria to Watson, August 9, 1951, *CSHLA*.
4. "As Watson wrote… ." Watson to Elizabeth Watson, July 14, 1951, *CSHLA*.
5. Watson JD, personal communication, 2012; Soderqvist, 2003.

Chapter 6

1. "Perutz had realized… ." Perutz, 2002, pp.189–191; Perutz MF. 1951. New X-ray evidence on the configuration of polypeptide chains. *Nature* **167:** 1053–1054.
2. "Watson's first impressions… ." Watson to Elizabeth Watson, September 1951, *CSHLA*.
3. "Watson's description… ." Watson to Elizabeth Watson, September 1951, *CSHLA*.
   "The membership… ." Perutz MF. 1970. Bragg, protein crystallography and the Cavendish Laboratory. *Acta Cryst* **26:** 183–185.
4. "Kalckar's enthusiastic letter… ." Kalckar to Lapp, October 5, 1951, *CSHLA*.
5. "Watson wrote to his sister… ." Watson to Elizabeth Watson, October 16, 1951, *CSHLA*; Watson to Elizabeth Watson, November 28, 1951, *CSHLA*.
6. "The letter from Luria… ." Luria to Watson, October 20, 1951, *CSHLA*; Matthews REF. 1989. Roy Markham: Pioneer in plant pathology. *Annu Rev Phytopathol* **27:** 13–22.
7. "My lodgings (digs) are adequate… ." Watson to Elizabeth Watson, October 9, 1951, *CSHLA*.
   "My digs situation… ." Watson to Elizabeth Watson, January 28, 1952 (This is misdated as "1951."), *CSHLA*.

8. "Despite the hardships… ." Watson to Elizabeth Watson, November 4, 1951, *CSHLA*.

Chapter 7

1. "Watson enthused… ." Watson to Delbrück, December 9, 1951, *CALTA*. "Crick also made… ." Crick, 1988, p. 75.
2. "Kendrew began working… ." Kendrew JC. 1964. Myoglobin and the structure of proteins. In *Nobel lectures, chemistry 1942–1962*, pp. 676–698. Elsevier, Amsterdam; Holmes KC. 2001. Sir John Cowdery Kendrew. March 24, 1917– August 23, 1997. *Biogr Mems Fell R Soc* **47**: 311–332.
3. "Alexander Todd was… ." Todd, 1973; Brown DM, Kornberg H. 2000. Alexander Robertus Todd, O.M., Baron Todd of Trumpington. October 2, 1907–January 10, 1997. *Biogr Mems Fell R Soc* **46**: 515–532.
4. "(*Left*) X-ray diffraction photograph… ." Bell FO. 1939. "X-ray and related studies of the structure of the proteins and nucleic acids." Ph.D. thesis, University of Leeds; Gosling R, personal communication, January 28, 2012.
5. "Astbury worked with… ." Bernal JD. 1963. William Thomas Astbury. 1898–1961. *Biogr Mems Fell R Soc* **9**: 1–35.
   "Florence Bell joined… ." Hall K. 2011. William Astbury and the biological significance of nucleic acids, 1938–1951. *Stud Hist Philos Biol Biomed Sci* **42**: 119–128.
6. "Watson wrote to his sister… ." Watson to Elizabeth Watson, November 14, 1951, *CSHLA*.
7. "The 'good crystalline DNA'… ." Wilkins, 2003, pp. 157 and 177.

Chapter 8

1. "This research was published… ." Crick is acknowledged in Bragg WL, Perutz MF. 1952. The external form of the haemoglobin molecule. I. *Acta Cryst* **5**: 277–283, p. 283. The other papers are: Bragg L, Perutz MF. 1952. The structure of haemoglobin. *Proc Roy Soc Lond Series A* **213**: 425–435; Bragg WL, Perutz MF. 1952. The external form of the haemoglobin molecule. II. *Acta Cryst* **5**: 323–328.
2. "The Physics Laboratory… ." Olby, 2009, p. 43.
3. "A. V. Hill, muscle physiologist… ." Katz B. 1978. Archibald Vivian Hill. *Biogr Mems Fell R Soc Lond* **24**: 71–149.
4. "Thomas Pigg Strangeways… ." Fell, 1962.
5. "Crick studied… ." Crick FHC, Hughes AFW. 1950. The physical properties of cytoplasm. A study by means of the magnetic particle method. Part I. Ex-

perimental. *Exp Cell Res* **1:** 37–80; Crick FHC. 1950. The physical properties of cytoplasm. A study by means of the magnetic particle method. Part II. Theoretical treatment. *Exp Cell Res* **1:** 505–533.

6. "A letter from A. V. Hill… ." A. V. Hill to Crick, March 11, 1949, *WL.*
"Crick worked with Perutz… ." Olby, 2009, pp. 94–96.

7. "Bragg regarded Crick… ." Bragg to A. V. Hill, January 18, 1952, quoted in Olby, 2009, p. 136.

Chapter 9

1. "Vladimir Vand was a Czech… ." *Physics Today*. July 1968, 115.
"Bill Cochran… ." Woolfson M. 2005. William Cochran. July 30, 1922–August 28, 1962. *Biogr Mems Fell R Soc* **51:** 67–85.
"Bill Cochran was reluctant… ." Cochran to Olby, July 19, 1968, quoted in Olby, 1994.

2. "Crick had married Doreen Dodd… ." Olby, 2009, pp. 54–55.

3. "Crick wrote later… ." Crick to Watson, March 31, 1966, *CSHLA.*

4. "According to the Oxford English Dictionary… ." "This I'm bound to say: four sweeter lovelier popsies, never blessed…" in Coyne, 1855.

5. "Odile described… ." Olby, 2009, p. 427.

6. "Cochran wrote later… ." Cochran W in Thomas and Phillips, 1990, p. 105.

7. "Later, Crick, writing… ." Crick to Watson, March 31, 1966, *WL.*

Chapter 10

1. "The day before… ." Watson to his parents, November 20, 1951, *CSHLA.*

2. "Watson's notorious… ." Raymond Gosling, BBC Radio 4, September 30, 2010.

3. "Shown above are pages… ." Rosalind Franklin's notebook, *CCAC.*

4. "Franklin's lack of enthusiasm… ." Franklin to Sayre, March 1, 1952, *ASM.*

5. "As late as 1955… ." Darlington CD. 1955. The chromosome as a physicochemical entity. *Nature* **176:** 1139–1144.

6. "A number of noted scientists… ." Werskey, 1978.

Chapter 11

1. "At the time of… ." Dodson G. 2002. Dorothy Mary Crowfoot Hodgkin, O.M. May 12, 1910–July 29, 1994. *Biogr Mems Fell R Soc* **48:** 179–219.

2. "Crick was indeed misled… ." Crick, 1988, p. 65.

3. "Bragg et al's… ." Bragg L, Kendrew JC, Perutz MF. 1950. Polypeptide chain configurations in crystalline proteins. *Proc Roy Soc Lond Series A* **203:** 321–357.

4. "At a July 1951 seminar… ." Crick, 1988, p. 50.
5. *The Nature of the Chemical Bond… .*" Pauling, 1939. The second edition used by Watson was published in 1940.
6. "This was Watson's first introduction… ." Watson to Elizabeth Watson, November 28, 1951, *CSHLA*.
7 "Watson told Delbrück… ." Watson to Delbrück, December 9, 1951, *CALTA*.
8. "Crick first met Kreisel… ." Olby, 2009, p. 58.

Chapter 12

1. "Kendrew had married Elizabeth… ." Holmes, 2001, p. 317.
2. "Watson refers to Sven… ." Olby, 1994, p. 336.
3. "The novel was *A Perch in Paradise*… ." Bullard, 1952; Bertrand Russell to Margaret Bullard, April 10, 1952, quoted in Blackwell K. 1995. Two days in the dictation of Bertrand Russell. *Russell: The journal of the Bertrand Russell Archives*, Summer 1995, 37–52, p. 44.
4. "The Rock of Gibraltar… ." Ellicot, 1975.
5. "Eprime Eshag was… ." Joshi H. 1998. Obituary: Eprime Eshag. *The Independent* (*London*), December 15, 1998.
6. "This is the first page of a memorandum… ." *WL*.
7. "Alexander Todd went to Frankfurt… ." Todd, 1973, p. 21.

Chapter 13

1. "Forty years later, Stokes recalled… ." Stokes AR. 1995. Why did we think DNA was helical? In Chomet, 1995, pp. 27–42.
2. "Raymond Gosling's account… ." Gosling R, personal communication, January 28, 2012.

Chapter 14

1. "We know now… ." Gann A, Witkowski JA. 2010. The lost correspondence of Francis Crick. *Nature* **467**: 519–524.
   "An exchange of letters… ." Wilkins to Crick, December 11, 1951, *CSHLA*. "Again from Wilkins to Crick… ." Wilkins to Crick, December 11, 1951, *CSHLA*.
   "On receipt… ." Crick to Wilkins, December 13, 1951, *CSHLA*.
2. "Watson is unfairly dismissive… ." Bragg WL, Nye JF. 1947. A dynamical model of a crystal structure. *Proc Roy Soc A* **190**: 474–481.
3. "A jig is used… ." Wilkins, 2003, pp. 175–176. The Merriam-Webster dictionary defines a jig as: "a device used to maintain mechanically the correct po-

sitional relationship between a piece of work and the tool or between parts of work during assembly."

4. "Crick's work on superhelical arrangements… ." Olby, 2009, pp. 144–145.

Chapter 15

1. "Letter from Watson to his sister… ." Watson to Elizabeth Watson, December 1952, *CSHLA*.
2. "All the pictures above… ." Edwards, 2000, pp. 374 and 388.
3. "Doris Lessing, Nobel Laureate… ." Lessing, 1997, p. 125.
4. "Writing to his parents… ." Watson to his parents, January 8, 1952 (misdated 1951); Watson to Elizabeth Watson, January 17, 1952, *CSHLA*.
5. "The Fellowship saga continues… ." Luria to Watson, March 5, 1952, *CSHLA*.

Chapter 16

1. "Watson had turned to TMV… ." Watson to Delbrück, May 20, 1952, *CALTA*.
2. "Bernal was one of the most outspoken… ." Brown, 2005; Bernal, 1939; Finney JL. 2007. *Journal of Physics: Conference Series* **57**: 40–52.
3. "Schramm was part… ." Macrakis K. 1993. The survival of basic biological research in National Socialist Germany. *J Hist Biol* **26**: 519–543; Lewis J. 2004. From virus research to molecular biology: Tobacco mosaic virus in Germany, 1936–1956. *J Hist Biol* **37**: 259–301, pp. 275–276.
4. "Geoffery Roughton's enthusiasm… ." Roughton to Bragg, August 5, 1951, Royal Institution of Great Britain; Gibson QH. 1973. Francis John Worsley Roughton. 1899–1972. *Biogr Mems Fell R Soc* **19**: 563–582; Deutsch T. 1995. Dr. Alice Roughton, *The Independent*, June 29, 1995.
5. "The Royal Society's… ." Astbury WT. 1953. Introduction. *Proc Roy Soc Lond Series B* **141**: 1–9, p. 5.

Chapter 17

1. "Initially named… ." Stoff, 2009.
2. "Letter from Ruth B. Shipley… ." Kahn J. 2011. The extraordinary Mrs. Shipley: How the United States controlled international travel before the age of terrorism. *Conn Law Rev* **43**: 819–888.
3. "The witholding of… ." *Los Angeles Examiner*, May 12, 1952; *The Times*, May 2, 1952.
4. "In a letter to his sister… ." Watson to Elizabeth Watson, April 3, 1952, *CSHLA*. "Left-leaning attitudes among scientists… ." The National Archives, United Kingdom. Wilkins' files references KV 2/3382 and KV 2/3383.

5. "François Jacob, then still a student… ." Jacob, 1988, p. 264.
6. "Having met at Cambridge… ." Pirie NW. 1973. Frederick Charles Bawden (1908–1972). *Biogr Mems Fell R Soc* **19**: 19–63; Pierpoint WS. 1999. Norman Wingate Pirie. July 1, 1907–March 29, 1997. *Biogr Mems Fell R Soc* **45**: 397–415; Jacob, 1988, p. 263.

Chapter 18

1. "Inspired by Avery's demonstration… ." Chargaff, 1978, pp. 85–86; Chargaff, 1963, p. 176.
2. "A table from Chargaff's 1950 paper… ." Chargaff, 1950. Chemical specificity of nucleic acids and mechanism of their enzymatic degradation. *Experientia* **6**: 201–209.
   "DNA now is a magic name… ." Chargaff, 1963, p. 162.
3. "Crick knew many scientists… ." Watson to Delbrück, December 9, 1951, *CALTA*.
4. "The Perfect Cosmological Principle… ." Bondi H. 2006. Thomas Gold. May 22, 1920–June 22, 2004. *Biogr Mems Fell R Soc* **52**: 117–135; Roxburgh IW. 2007. Hermann Bondi. November 1, 1919–September 10, 2005. *Biogr Mems Fell R Soc* **53**: 45–61.
5. "Pascual Jordan, the theoretical physicist… ." Quoted in Beyler RH. 2007. Exporting the quantum revolution: Pascual Jordan's biophysical initiatives. In Hoffmann et al., 2007; Beyler RH. 1996. Targeting the organism: The scientific and cultural context of Pascual Jordan's quantum biology, 1932–1947. *Isis* **87**: 248–273.
6. "On the left is the article… ." Pauling L, Delbrück M. 1940. The nature of the intermolecular forces operative in biological processes. *Science* **92**: 77–79.
7. "We do not have any documentation… ." Griffith to Crick, March 2, 1953, *CSHLA*.
8. "Chargaff wrote a vivid account… ." Chargaff, 1978, p. 101; Judson, 1996, p. 119.

Chapter 19

1. "In 1974, Chargaff… ." Chargaff E. 1974. Building the Tower of Babble. *Nature* **248**: 776–779.
   "This example… ." Watson to Delbrück, January 4, 1954, *CALTA*.
2. "Born in Moscow… ." Roman H. 1980. Boris Ephrussi. *Ann Rev Genet* **14**: 447–450; Ravin AW. 1968. Harriett Ephrussi–Taylor. April 10, 1918–March 30, 1968. *Genetics* (suppl.) **60**: p. 24; Berg and Singer, 2003, pp. 101–113.
3. "After being denied a passport… ." Hager, 1995, pp. 406–407.
4. "Wilkins describes the reason… ." Wilkins, 2003, pp. 181 and 187.
5. "Watson wrote to the Cricks… ." Watson to Crick, August 11, 1952, *WL*.

6. "Born into a family of Bostonian bluebloods… ." Alberty RA, di Cera E. 2003. Jeffries Wyman 1901–1994. *Biog Mem Natl Acad Sci* **83:** 2–17; Wyman AC, 2010.

7. "Watson reported to the Cricks… ." Watson to Crick, August 11, 1952, *CSHLA.*

8. "The hostess was actually the Baroness *Edouard* de Rothschild… ." Morton, 1962.

9. "As indicated by this postcard… ." Watson to Elizabeth Watson, August 26, 1952, *CSHLA.*

10. "At this time, Spiegelman… ." Garfield E. 1983. They stand on the shoulders of giants: Sol Spiegelman, a pioneer in molecular biology. *Essays of an information scientist* **6:** 164–171.

Chapter 20

1. "Hayes studied in Ireland… ." Broda P, Holloway B. 1996. William Hayes. January 18, 1913–January 7, 1994. *Biogr Mems Fell R Soc* **42:** 172–189.

2. "Norton Zinder, then a graduate student… ." Zinder N. 1992. Forty years ago: The discovery of bacterial transduction. *Genetics* **132:** 291–294, p. 293.

3. "The rabbinical complexity… ." Ephrussi B, Leopold U, Watson JD, Weigle J. 1953. Terminology in bacterial genetics. *Nature* **171:** 701; Watson, 2001. *Genes, girls and Gamow*, p. 12.

4. "As Watson wrote to his sister… ." Watson to Elizabeth Watson, October 27, 1952, *CSHLA.*

5. "Wilkins later wrote to Crick… ." Wilkins to Crick, undated but early 1952, *CSHLA.*

6. "Watson in a letter… ." Watson to Delbrück, May 20, 1952, *CALTA.*

7. "Crick, Pauling, and coiled-coils… ." Crick FHC. 1952. Is α-keratin a coiled coil? *Nature* **170:** 882–883; Pauling L, Corey RB. 1953. Compound helical configurations of polypeptide chains: Structure of proteins of the α-keratin type. *Nature* **171:** 59–61; Peter Pauling to Linus Pauling, January 13, 1953. "The matter was further complicated… ." Linus Pauling to Perutz, March 29, 1953, *OSUSC.* "Crick replied… ." Crick to Linus Pauling, April 14, 1953, *OSUSC.*

8. "In 1949, Irving Langmuir… ." Hauptman HA. 1998. David Harker. 1906–1991. *Bio Mem Natl Acad Sci* **74:** 126–143.

9. "Hayes' laboratory… ." Broda and Holloway, 1996. See note 1 above.

10. "Franklin's move to Birkbeck… ." Maddox, 2002, p. 183; Franklin to Sayre, March 1, 1952; Sayre to Franklin, March 8, 1952; Franklin to Sayre, June 2, 1952, *ASM*; Franklin to Bernal, June 19, 1952, *CCAC.*

Chapter 21

1. "Clare College, the second oldest… ." Eden, 1998.
   "In a letter from Watson… ." Watson to Elizabeth Watson, October 8, 1952, *CSHLA.*
2. "Denys Wilkinson… ." Transcript of interview of Sir Denys Wilkinson by CJ Meyer, 1964, Niels Bohr Library & Archives, American Institute of Physics, College Park, Maryland. http://www.aip.org/history/ohilist/876.html.
3. "Nicholas G. L. Hammond was a scholar… ." Clogg R. 2001. Nicholas Hammond. *The Guardian*, Wednesday April 4, 2001.
4. "The local doctor Watson visited… ." Watson JD, personal communication; Monk, 1990, pp. 576–579.
5. "Writing to Delbrück… ." Watson to Delbrück, December 9, 1951, *CALTA.*
6. "Watson's French lessons… ." Watson to Elizabeth Watson, October 8, 1952, *CSHLA.*
   "Mrs. Camille Prior… ." Roberts, 2010.
7. "Pauling began thinking intensively… ." Linus Pauling, notebook, November 26, 1952, *OSUSC.*
8. "In reply to this letter… ." Peter Pauling to Linus Pauling, January 13, 1953, *OSUSC.*
9. "In addition to his son… ." Linus Pauling to Randall, December 31, 1952, *OSUSC.*

Chapter 22

1. "Pasadena, the home… ." Hager, 1995.
2. "Writing of her imminent move… ." Franklin to Adrienne Weill, March 10, 1953, *ASM*; Franklin to Ann Sayre, December 17, 1953.
3. "Randall reminded Franklin… ." Randall to Franklin, April 17, 1953, *CCAC.*
4. "Verner Schomaker, pictured on page 225… ." Quoted in Dunitz J. 1997. Linus Pauling. February 1901–August 19, 1994. *Bio Mem Natl Acad Sci* **71:** 221–261, p. 243.
5. "It was two years… ." Hager, 1995, pp. 449–458 and 546–554.

Chapter 23

1. "Franklin and helical DNA… ." Klug A. 1968. Rosalind Franklin and the discovery of the structure of DNA. *Nature* **219:** 808–844; Olby, 1994, pp. 370–376; Crick to Wilkins, June 5, 1953, *CSHLA.*
2. "Wilson joined Wilkins… ." Wilkins, 2003, p. 196.
3. "Notebook recording DNA… ." *KCLA.*
4. "An example of the kind… ." Wilkins to Crick, undated but early 1952, *WL.*

5. "R. D. B. 'Bruce' Fraser's model of DNA... ." Fraser's unpublished manuscript was printed 50 years later. Fraser RDB. 2004. The structure of deoxyribose nucleic acid. *J Struct Biol* **145**: 184–186.

6. "The Famous Photograph 51... ." Gosling R, personal communication, January 28, 2012; Wilkins, 2003, pp. 197–198.

7. "Peter Pauling tells his father... ." Peter Pauling to Linus Pauling, January 13, 1953, *OSUSC*.

8. "Wilkins finally found a new flat... ." Wilkins to Crick, June 3 1953, *CSHLA*.

9. "In a letter from Wilkins to Crick... ." Wilkins to Crick, probably February, 1953, *CSHLA*.

Chapter 24

1. "Watson had written... ." Watson to Elizabeth Watson, December 11, 1952, *CSHLA*.

2. "The 'most beautiful male' in Cambridge... ." Bertrand Fourcade, personal communication, March 28, 2012; Russell J. 1987. Xavier Fourcade dead at 60; Dealer in contemporary art. *The New York Times*, April 29, 1987; Vogel C. 1992. Vincent Fourcade, 58, decorator known for his ornate interiors. *The New York Times*, December 25, 1992.

3. "The Rolls-Royce... ." Watson JD, personal communication; Crossette B. 2003. Geoffrey Bawa, 83, architect who blended Asian styles, dies. *The New York Times*, May 31, 2003.

4. "'Their host' was Albert Edward Richardson... ." Watson JD, personal communication; "Sir Albert Richardson: The Complete Georgian." *The Times* obituary, February 4, 1964.

5. "Victor Rothschild was Nathaniel Mayer Victor Rothschild... ." Reeve S. 1994. Nathaniel Mayer Victor Rothschild, G. B. E., G. M. Third Baron Rothschild. October 31, 1910–March 20, 1990. *Biogr Mems Fell R Soc* **39**: 364–380; Andrew, 2009, pp. 269–270 and 377.

6. "Wilkins reflected on this key episode... ." Wilkins, 2003, pp. 205–206.

Chapter 25

1. "Watson's enthusiasm for tennis... ." Watson to Elizabeth Watson, April 27, 1952; Watson to Elizabeth Watson, July 8, 1952, *CSHLA*.

2. "The Rex cinema... ." Binder, 2011.

3. "This 1933 Czech film... ." Shearer, 2010.

4. "The front page of the report... ." Perutz MF, Wilkins MHF, Watson JD. 1969. DNA helix. *Science* **164**: 1537–1539.

5. "The significance of… ." Klug A. 1968. Rosalind Franklin and the discovery of the structure of DNA. *Nature* **219**: 808–844; Olby, 2009, pp. 161–163.

6. "This work, begun by J. M. Gulland… ." Haworth RD. 1948. John Masson Gulland 1898–1947. *Obit Not Fell R Soc* **6**: 67–82; Harding SE, Winzor DJ. 2010. James Michael Creeth (1924–2010). *Macromol Biosci* **10**: 696–699; Manchester KL. 1995. Did a tragic accident delay the discovery of the double helical structure of DNA? *Trends Biochem Sci* **20**: 126–128; Trench AC, Wilson GRS. 1948. Report on the derailment which occurred on October 26, 1947, at Goswick on the London and North Eastern Railway, His Majesty's Stationery Office, London.

Chapter 26

1. "Despite having reservations… ." Delbrück to E.B. Wilson, February 25, 1953, *CSHLA*.

   "Delbrück sent Watson a copy… ." Delbrück to Watson, February 25, 1953, *CSHLA*.

2. "There is a minor mystery… ." Davidson, 1950, p. 6.

3. "Watson mentions in the caption… ." Pauling, Solvay Conference Notebook, April 8, 1953, *OSUSC*.

4. "Crick did not recall… ." Crick, 1988, p. 77.

Chapter 27

1. "The oldest surving bridge… ." Ray, 1994.

2. "The second oldest building… ." Ray, 1994.

3. "By the early 1950s, Heffer's… ." Wilson, 2010.

4. "The original DNA model… ." MRC Laboratory of Molecular Biology, Cambridge.

5. "The DNA helix is right-handed… ." Kamminga and de Chadarevian, 2002.

6. "The Cricks' imminent exile… ." Crick to David Harker, January 21, 1953, *WL*; Olby, 2009, pp. 119–129.

7. "On March 7, 1953, Wilkins wrote to Crick… ." Wilkins to Crick, March 7, 1953, *WL*.

8. "Crick's response to Wilkins' letter… ." Judson, 1996, p. 151; Judson's interview with Crick, September 10, 1975.

Chapter 28

1. "Wilkins described his reaction… ." Wilkins, 2003, p. 212.

2. "Jerry Donohue appears to have had mixed feelings… ." Donohue J. 1969.

Fourier analysis and the structure of DNA. *Science* **165**: 1091–1096; Crick FHC. 1970. DNA: Test of structure? *Science* **167**: 1694.

3. "Wilkins remembered the scene differently… ." Wilkins, 2003, p. 214.

4. "Franklin's notes shown here… ." Franklin's notebook, February 23, 1953, *CCAC*.

5. "Unbeknown to Watson, Crick, and Wilkins… ." Franklin and Gosling draft, March 17, 1953, *CCAC*.

6. "Pauling had written to Peter… ." Linus Pauling to Peter Pauling, February 18, 1953, *OSUSC*.

7. "Robert Corey was Pauling's right-hand… ." Marsh RE. 1997. Robert Brainard Corey. 1897–1971. *Bio Mem Natl Acad Sci* **72**: 50–68.
"Verner Schomaker was a chemist at Caltech… ." Trueblood K. 1997. Verner Schomaker (1914–1997). *J Appl Cryst* **30**: 526.

8. "This picture was taken by… ." Antony Barrington Brown Obituary. *The Telegraph* (*London*), February 14, 2012; de Chadarevian S. 2003. Portrait of a discovery: Watson, Crick, and the double helix. *Isis* **94**: 90–105.

9. "Surely Watson's reference… ." Todd, 1973, p. 89.

10. "As shown in this diagram… ." Delbrück to Watson, May 12, 1953, *CSHLA*; Delbrück M, Stent G. 1957. On the mechanism of DNA replication. In McElroy and Glass, 1957, pp. 699–736; Ray Gosling interviewed by Jane Callander, June 24, 1985, cited by Maddox, 2002; Wang JC. 1971. Interaction between DNA and an *Escherichia coli* protein ω. *J Mol Biol* **55**: 523–526.

Chapter 29

1. "In his letter to Delbrück… ." Watson to Delbrück, March 12, 1953, *CALTA*.

2. "Funded by the industrialist… ." Marage and Wallenborn, 1999.

3. "Watson described these… ." Watson to Delbrück, March 22, 1953, *CALTA*.

4. "This paper was submitted… ." Wyatt GR, Cohen SS. 1953. The bases of the nucleic acids of some bacterial and animal viruses: The occurrence of 5-hydroxy-cytosine. *Biochem J* **55**: 774–782.

5. "On this and the following pages… ." Crick to Wilkins, March 17, 1953, *CSHLA*; Wilkins to Crick, March 18, 1953, quoted in Olby, 1994, p. 417; Wilkins to Crick, March 23, 1953, *WL*.

6. "Gerald Pomerat, assistant director… ." Witkowski JA. 2002. Mad hatters at the DNA tea party. *Nature* **415**: 473–474.

7. "Pauling's graceful acceptance… ." Pauling to Ava Pauling, April 6, 1953; Pauling to Delbrück, April 20, 1953, *OSUSC*.

8. "In a similar mood… ." Watson to Delbrück, March 22, 1953, *CALTA*.

Epilogue

1. Kennedy EP. 1996. Herman Moritz Kalckar. 1908–1991. *Bio Mem Natl Acad Sci* **69:** 148–164.
2. Holmes KC. 2001. Sir John Cowdery Kendrew. March 24, 1917–August 23, 1997. *Biogr Mems Fell R Soc* **47:** 311–332; Kendrew JC. 1980. The European molecular biology laboratory. *Endeavour* **4:** 166–170; Krige J. 2002. The birth of EMBO and the difficult road to EMBL. *Stud Hist Phil Biol & Biomed Sci* **33:** 547–564.
3. Ferry, 2007; Perutz, 2002. Perutz made a definitive collection of his papers on haemoglobin, together with essays on his research and the attendant controversies, in Perutz, 1997.
4. Phillips D. 1979. William Lawrence Bragg. March 31, 1890–July 1, 1971. *Biogr Mems Fell R Soc* **25:** 74–143.
5. Huxley H, personal communication, 2012.
6. Ridley, 2006; Olby, 2009.
7. Wilkins, 2003; Burhop EHS. 1971. The British Society for Social Responsibility in Science. *Phys Educ* **6:** 140–142.
8. Hager, 1995.
9. Maddox, 2002; Glynn, 2012; Wolstenholme GEW, Millar ECP, eds. 1957. *The nature of viruses CIBA foundation symposium*. J. & A. Churchill Ltd, London. Harrington's remark is on page 285.

The Nobel Prize

This is an abridged version of chapter 10 "Manners Appropriate for a Nobel Prize" in *Avoid boring people: Lessons from a life in science*, by James D. Watson (2007), published by Alfred A. Knopf, a division of Random House, Inc., and Oxford University Press.

1. Rocca, 2007.
2. "Charles Huggins was a cancer researcher... ." Forster RE. 1999. Charles Brenton Huggins (September 22, 1901–January 12, 1997). *Proc Am Philos Soc* **143:** 325–331; Dulbecco R. 1976. Francis Peyton Rous. October 5, 1879–February 16, 1970. *Bio Mem Natl Acad Sci* **48:** 275–306.
3. "The receipt for... ." *CSHLA.*
4. "Crick wrote to Watson... ." Crick to Watson, October 30, 1962, *CSHLA.*
5. "Nathan M. Pusey was... ." Yarrow AL. 2001 Nathan Pusey, Harvard President through growth and turmoil alike, dies at 94. *The New York Times*, November 15, 2001.
6. "Henry Stuart Hughes was an historian... ." Mosteller DP. 1999. Former his-

tory prof., activist Hughes dies at 83. *The Harvard Crimson*, Monday, October 25, 1999.

7. "At the Laboratory for Molecular Biology… ." Finch, 2008.

8. "George Gamow was a physicist… ." Watson, 2001. *Genes, girls and Gamow*; Gamow G. 1970. *My world line: An informal autobiography*. Viking, New York; Hufbauer K. (forthcoming) George Gamow. 1904–1968. *Bio Mem Natl Acad Sci*.

9. "The congratulatory telegram… ." *CSHLA*.

10. "Kai Falkman went on… ." Watson JD, personal communication, 2012.

11. "The first page… ." *CSHLA*.

12. "James Graham Parsons was a career diplomat… ." Pace E. 1991. J. Graham Parsons is dead at 83; former envoy to Laos and Sweden. *The New York Times*, October 22, 1991.

13. "John Franklin Enders shared… ." Weller TH, Robbins FC. 1991. John Franklin Enders (1897–1985). *Bio Mem Natl Acad Sci* **60:** 47–65.

14. "St. Lucia girls… ." Eriksson SA. 2002. Christmas traditions and performance rituals: A look at Christmas celebrations in a Nordic context. *Applied Theatre Researcher* 3.

15. "Bo Beskow was… ." Beskow, 1969.

Photo Credits

64, right, with permission from the American Society for Microbiology, Anne Sayre Collection; p. 65, courtesy of Jenifer Glynn; p. 66, http://rackandruin.blogspot.com/2009/02/foggy-day-in-london-town.html; p. 67, with permission from the American Society for Microbiology, Anne Sayre Collection; p. 68, with permission from King's College London Archives; p. 69, ©BBC.

Chapter 11: p. 71, Photograph by Gordon Cox, courtesy of Judith Howard, from *Dorothy Hodgkin*, by Georgina Ferry, ©Cold Spring Harbor Laboratory Press; p. 72, with permission from News International Trading Ltd.; p. 73, Bragg et al. (1950), *Proc R Soc Lond A Math Phys Sci 203:* 321–357, with permission from The Royal Society; p. 75, with permission from the Ava Helen and Linus Pauling Papers, Special Collections, Oregon State University; p. 76, top, with permission from ©Oxfordshire County Council, Oxfordshire History Centre; p. 76, bottom, ©Rockefeller University Press, originally published in (2006), *J Exp Med 203:* 809–818, with permission; p. 77, top, ©Prabhu B. Doss; p. 77, bottom, with permission from University of California, San Diego, Mandeville Special Collections Library.

Chapter 12: p. 79, With permission from The British Library Board, *The Times*, October 20, 1951; p. 80, left, courtesy of Bjørn Pedersen; p. 81, Furberg (1949), *Nature 164:* 22, with permission from Macmillan; p. 82, H. Hamilton, publisher, with permission from Penguin Group; p. 83, courtesy of Herbert Gutfreund; p. 85, with permission from Wellcome Library, London; p. 87, courtesy of Bruce Fraser.

Chapter 13: pp. 89–90, left, With permission from King's College London Archives; p. 90, right, ©NRM/Science and Society Picture Library; p. 92, ©Science and Society/SuperStock; p. 93, with permission from King's College London Archives.

Chapter 14: p. 95, Photograph by Norton Hintz, courtesy of AIP Emilio Segre Visual Archives, Hintz Collection; pp. 96–98, courtesy of Cold Spring Harbor Laboratory Archives; p. 99, Bragg and Nye (1947), *Proc R Soc Lond A 190:* 474–481, with permission from The Royal Society.

Chapter 15: p. 103, top, ©Picture Post/Malcolm Dunbar/Getty Images; p. 103, bottom left, with permission from www. britishcouncil.org/film; p. 103, bottom right, source unknown; p. 104, courtesy of the James D. Watson Collection, Cold Spring Harbor Laboratory Archives; p.

105, top, http://www. kintyreonrecord.co.uk; p. 105, bottom, with permission from Eleanor and Clyde Moore (www.PhotosByEleanor. com); p. 106, left, portrait by Percy Wyndham Lewis, with kind permission from Avrion Mitchison; p. 106, middle, Percy Wyndham Lewis, (Lady) Naomi Mitchison, Scottish National Portrait Gallery, purchased with assistance from the Art Fund and the Patrons of the National Galleries of Scotland 2003; p. 106, right, W.P. Stern Rare Books; p. 107, with permission from Neil Mitchison; p. 108, top, ©Anne Burgess; p. 108, bottom, courtesy of Avrion Mitchison; p. 109, courtesy of Jan Witkowski.

Chapter 16: p. 112, ©IUCr; p. 113, courtesy of the Sydney Brenner Collection, Cold Spring Harbor Laboratory Archives; p. 114, with permission from Schramm (1947), *Z Naturforsch 2b:* 112–121; p. 115, 116, left, with permission from *J Gen Physiol 25:* 147–165, ©1941 Rockefeller University Press; p. 116, right, Jones R.V. (1978), *Most secret war: British scientific intelligence 1939–1945.* H. Hamilton, with permission from the estate of R.V. Jones; p. 117, with permission from Royal Society of Chemistry.

Chapter 17: p. 119, http://www.postcardpost.com/enell.htm, with permission from Larry Myers; p. 120, with permission from the Ava Helen and Linus Pauling Papers, Special Collections, Oregon State University; p. 121, left, *Los Angeles Examiner*, May 2, 1952, with permission from Hearst Newspapers; p. 121, right, *The Times*, May 2, 1952, with permission from NI Syndication; p. 122, with permission from the Society for General Microbiology; p. 123, with permission from The National Archives; p. 124, top and bottom right, courtesy of Cold Spring Harbor Symposia on Quantitative Biology Collection, Cold Spring Harbor Laboratory Archives; p. 124, bottom left, courtesy of the James D. Watson Collection, Cold Spring Harbor Laboratory Archives; p. 125, left, *J Gen Microbiol* (1972), *72(1)*, with permission; p. 125, right, with permission from Lebrecht Music & Arts; p. 126, courtesy of the James D. Watson Collection, Cold Spring Harbor Laboratory Archives.

Chapter 18: pp. 127–129, Watson (1954), *Biochim Biophys Acta 13:* 10–19, with permission from Elsevier; p. 130, courtesy of the Cold Spring Harbor Laboratory Library Archives; p. 131, top, Chargaff (1950), *Experientia 6:* 201–209, with kind permission from Springer Science + Business Media B.V.; p. 131, bottom, copyright unknown; p. 132, with permission from John Wiley & Sons; p. 133, top,

©Fotothek, SLUB, Dresden; **p. 133, bottom,** with permission from American Association for the Advancement of Science; **p. 134,** courtesy of the Sydney Brenner Collection, Cold Spring Harbor Laboratory Archives; **p. 135,** courtesy of the James D. Watson Collection, Cold Spring Harbor Laboratory Archives.

Chapter 19: p. 137, Courtesy of Jan Witkowski; **p. 138,** ©Esther M. Zimmer Lederberg Memorial Website (www.estherlederberg.com); **p. 139,** with permission from the Ava Helen and Linus Pauling Papers, Special Collections, Oregon State University; **p. 140,** Wikipedia; **p. 141, top,** courtesy of the James D. Watson Collection, Cold Spring Harbor Laboratory Archives; **p. 141, bottom,** Weyman and Gill (1987), *Annu Rev Biophys Biophys Chem 16:* 1–24, with permission from Annual Reviews; **p. 142,** with permission from the Ava Helen and Linus Pauling Papers, Special Collections, Oregon State University; **p. 143,** with permission from INFA; **p. 144,** courtesy of the James D. Watson Collection, Cold Spring Harbor Laboratory Archives; **p. 145, top,** ©Esther M. Zimmer Lederberg Memorial Website (www. estherlederberg. com); **p. 145, bottom,** courtesy of the James D. Watson Collection, Cold Spring Harbor Laboratory Archives.

Chapter 20: p. 147, ©Esther M. Zimmer Lederberg Memorial Website (www.estherlederberg.com); **p. 148.** courtesy of the Cold Spring Harbor Laboratory Library Archives; **p. 149.** Ephrussi B et al. (1953), *Nature 171:* 701, with permission from Macmillan; **p. 150,** ©Esther M. Zimmer Lederberg Memorial Website (www. estherlederberg.com); **p. 152,** Pauling and Corey (1953), *Nature 171:* 59–61, with permission from Macmillan; **p. 154,** courtesy of Jan Witkowski; **p. 155,** courtesy of Jenifer Glynn.

Chapter 21: p. 157, With permission from www.cambridge2000. com; **p. 158,** courtesy of Angelo Nardoni (Flickr); **p. 159, top,** kindly provided by Stephen Harris; **p. 159, bottom,** courtesy of Stuart Williams (Flickr); **pp. 160–161,** courtesy of Herbert Gutfreund; **pp. 164–166,** with permission from the Ava Helen and Linus Pauling Papers, Special Collections, Oregon State University; **p. 167,** courtesy of Hugh Huxley.

Chapter 22: p. 169, With permission from the Ava Helen and Linus Pauling Papers, Special Collections, Oregon State University; **p. 170,** Churchill Archives Centre, The Papers of Rosalind Franklin, FRKN 1/2; **pp. 172, 173,** with permission from the Ava Helen and Linus Pauling Papers, Special Collections, Oregon State University.

Chapter 23: p. 177, ©Rachel Glaeser/American Society for Microbiology; **p. 179,** courtesy of Raymond Gosling; **p. 180, left,** with permission from Wellcome Library, London; **p. 180, right,** with permission from King's College London Archives; **p. 181, top,** ©University of Dundee Archive Services, with permission; **p. 181, bottom,** courtesy of Bruce Fraser; **p. 182,** with permission from King's College London Archives; **p. 183,** with permission from the Ava Helen and Linus Pauling Papers, Special Collections, Oregon State University.

Chapter 24: p. 185, left, With permission from University of California, San Diego, Mandeville Special Collections Library; **p. 185, right,** with permission from Cambridge University Archives; **p. 186,** courtesy of Herbert Gutfreund; **p. 187,** photograph by Pierre Boulat, with permission from Annie Boulat; **p. 188,** with permission from Beds & Luton Archives Service; **p. 189,** National Portrait Gallery, London, with permission; **p. 190,** courtesy of Miriam Murphy.

Chapter 25: p. 193, With permission from Cambridge Newspapers Ltd.; **p. 194, left,** Wikipedia; **p. 194, right,** ©1950 Springer, with permission; **p. 195, top,** reproduced, with permission, from the Medical Research Council; **p. 195, bottom,** with permission from King's College London Archives; **p. 197, top left,** with permission from Harding and Winzor (2010), James Michael Creeth (1924−2010). *The Biochemist 32:* 44–45; **p. 197, top right,** courtesy of Steve Harding, University of Nottingham; **p. 197, bottom left,** photograph from *The Illustrated London News*; **p. 197, bottom right,** from Haworth R.D. (1948), John Masson Gulland 1898−1947. *Obituary Notices Fell R Soc 6:* 67−82.

Chapter 26: pp. 201−202, Courtesy of the James D. Watson Collection, Cold Spring Harbor Laboratory Archives; **p. 203,** ©1950 Springer, with permission; **pp. 205,** with permission from the Ava Helen and Linus Pauling Papers, Special Collections, Oregon State University; **p. 207,** courtesy of the James D. Watson Collection, Cold Spring Harbor Laboratory Archives; **p. 209,** Wikipedia.

Chapter 27: p. 211, ©A. Barrington Brown/Photo Researchers, Inc.; **p. 212,** ©Free Stock Photos; **p. 213, top,** ©txllxt, from panoramio. com, with permission; **p. 213, bottom,** flickr.com.photos; **p. 214,** ©MRC Laboratory of Molecular Biology; **pp. 216, 218,** with permission from Wellcome Library, London.

Chapter 28: **p. 223,** Courtesy of Jenifer Glynn; **p. 225,** courtesy of the Archives, California Institute of Technology; **p. 226,** ©A. Barrington Brown/Photo Researchers, Inc.

Chapter 29: **p. 229,** With permission from the Ava Helen and Linus Pauling Papers, Special Collections, Oregon State University; **pp. 230–231,** ©Biochemical Society, with permission; **pp. 233–235,** courtesy of the Sydney Brenner Collection, Cold Spring Harbor Laboratory Archives; **p. 236, top,** courtesy of Rockefeller Archive Center; **p. 236, bottom,** ©A. Barrington Brown/Photo Researchers, Inc.; **p. 237,** with permission from the Ava Helen and Linus Pauling Papers, Special Collections, Oregon State University; **p. 238,** from livejournal.com; **p. 240,** courtesy of Don Caspar.

The Nobel Prize: **p. 241, bottom,** Getty Images/AFP; **p. 242,** courtesy of the James D. Watson Collection, Cold Spring Harbor Laboratory Archives; **p. 243, top right,** with permission from CSU Archives/Everett Collection; **p. 243, bottom left and right,** courtesy of the James D. Watson Collection, Cold Spring Harbor Laboratory Archives; **p. 245, top left and right,** Hans Boye/MRC Laboratory of Molecular Biology; **p. 245, bottom left,** http://quages.com/others/

Lev-Landau.html; **p. 245, bottom right,** courtesy of Cold Spring Harbor Laboratory Library Archives; **p. 246,** Harvard University; **pp. 247–249,** courtesy of the James D. Watson Collection, Cold Spring Harbor Laboratory Archives; **p. 250, left,** Jan Ehnemark/Kamera Bild, Stockholm; **p. 250, right,** Pool/Scanpix Sweden/Sipa USA; **p. 251, top,** Scanpix Sweden/Sipa USA; **p. 251, bottom,** Lennart Edling/Kamera Bild, Stockholm; **p. 252,** Lennart Edling/Scanpix Sweden/Sipa USA; **p. 253,** courtesy of the James D. Watson Collection, Cold Spring Harbor Laboratory Archives; **p. 254,** Ragnhild Haarsta/SVD/Scanpix Sweden/Sipa USA; **p. 255,** courtesy of the James D. Watson Collection, Cold Spring Harbor Laboratory Archives; **p. 256,** Jan Delden/Scanpix Sweden/Sipa USA.

Appendices: **pp. 258–264,** Courtesy of Michael Crick; **pp. 265–267, 274,** courtesy of the James D. Watson Collection, Cold Spring Harbor Laboratory Archives; **p. 279,** courtesy of National Academy of Sciences Archives; **pp. 288, 289, 291–294, 296–298,** courtesy of the James D. Watson Collection, Cold Spring Harbor Laboratory Archives; **pp. 304–305,** with permission from the American Association for the Advancement of Science; **p. 306,** from Perutz MF (1969), *Science 164:* 1538, courtesy of Vivien Perutz.

Index

Page references in italics refer to information found in the annotations and figure legends.